Chris Rauwendaal

SPC
Statistical Process Control in Extrusion

Chris Rauwendaal

SPC

Statistical Process Control in Extrusion

HANSER

Hanser Publishers, Munich Vienna New York Barcelona

The Author:
Dr. Chris Rauwendaal, Rauwendaal Extrusion Engineering, Inc.,
12693 Roble Veneno Lane, Los Altos Hills, CA 94022, USA

Library of Congress Cataloging-in-Publication Data
Rauwendaal, Chris:
SPC-statistical process control in extrusion / Chris Rauwendaal.
p. cm.
Includes index.
ISBN 3-446-16244-5
1. Plastics – Extrusion. 2. Chemical process control – Statistical methods. I. Title.
TP1175.E9R38 1993
668.4′13–dc20

Die Deutsche Bibliothek – CIP-Einheitsaufnahme
Rauwendaal, Chris:
SPC – statistical process control in extrusion / Chris Rauwendaal. – Munich; Vienna; New York; Barcelona: Hanser; 1993
ISBN 3-446-16244-5

ISBN 3-446-16244-5 Carl Hanser Verlag, Munich Vienna New York Barcelona

Typeset and printed in Germany by Druckerei Appl, Wemding
Bound in Germany by Sellier, Freising

To Sietske, Randy-Roel, Lisette and Yolanda

PREFACE

Statistical process control, or SPC, has enjoyed a kind of renaissance in the 1990's. Originally developed in the 1920's by the famous American statistician Dr. Walter A. Shewhart, SPC was used in the first half of the century. However, lack of competition and complacency led to a declining use of SPC in American industry. From the 1950's to the 1980's the majority of the U.S. companies were not using SPC methods in their manufacturing operations. At the same time, Japanese companies readily accepted and implemented SPC under the tutelage of Dr. W. Edward Deming, a former colleague of Dr. Shewhart. There is no doubt that the widespread use of SPC by Japanese companies has contributed to their improved competitive position in the world market.

Exposure to increased competitive pressures has gradually led to the realization that SPC is an indispensable tool for the manufacturing industry. Some companies implement SPC because they understand that SPC is one of the most powerful tools in achieving improved productivity as well as product quality. Other companies introduce SPC because their customers require SPC data to convince themselves that the supplier can make the product in question to the required specifications reliably and consistently.

Whatever the motives, the 1990's have become a period of increased use of SPC in American industry. As a result, many companies have introduced training programs to teach their employees the basics of SPC. These training programs are frequently taught by statisticians who know little about the actual manufacturing processes used in the company. This can make it difficult to see how SPC can be used for a particular manufacturing process such as extrusion. If SPC is to be implemented successfully, training in the principles of statistical process control should be combined with training in the appropriate process technology.

This book, therefore, combines the principles of SPC with the main elements of extrusion technology. The aim of this book is to combine the benefits of proper implementation of SPC with the benefits of improved extrusion technology. The first four chapters deal with the basic principles of SPC: introduction, data collection and analysis, control charts, and process capability. Chapter five covers general aspects of measurements as well as specific measurements in extrusion. Chapter six deals with basic extrusion technology, followed by efficient machine design and process operation. The latter allows implementation of machine and process improvements that are not necessarily obvious from just applying SPC.

The final chapter covers advanced statistical techniques, such as ANOVA, design of experiments, regression analysis, response surface methodology, evolutionary operation, and the Taguchi method. This is the only chapter that requires a reasonable level of mathematical skill. The level of mathematical complexity of the first six chapters is kept to a minimum to make the book accessible and useful to technicians and operators as well as process engineers and production supervisors. A companion video training program "Statistical Process Control in Extrusion" is available through the Society of Plastics Engineers, Inc.

The author would like to thank Dr. E. Immergut for his encouragement to write this book. He would like to thank the following persons for useful contributions: Mr. Stan Vandercook, Mr. Peter Galuszka, Mr. Manuel Nunes, Dr. Kenny Leong, and Mr. Jack Contessa. Lastly, the author would like to thank his family for their patience and support in these time consuming projects.

Los Altos Hills, California, June 1992
Chris Rauwendaal

CONTENTS

1 INTRODUCTION TO STATISTICAL PROCESS CONTROL

1.1 INTRODUCTION

What is Statistical Process Control or SPC?
SPC simply means the use of statistical methods to monitor and control a process.

SPC can be applied to many different processes. Not only manufacturing processes, such as the production of electronic chips, batteries, tires, plastic bottles, etc., but also tasks performed in the service industries, such as filling out claim forms, diagnosis of patients by a medical doctor, processing of tax refunds by the IRS, processing of mail by the Postal Office, etc.

Statistics deals with the collection, organization, analysis, and interpretation of numerical data. An integral part of statistics is mathematics; this can make statistics rather imposing and unappetizing. However, the basic mathematical tools necessary to do SPC are straight forward and do not need to be intimidating at all. In the following sections, mathematics will be kept to a minimum to make the material accessible to as many people as possible. For those interested in more mathematical details, some of the appendices and chapter 7 will cover some more advanced mathematics.

The goals of SPC are to improve and ensure quality, and, thus, reduce process cost due to waste as a result of rejects. Quality is determined by how much the characteristic of a product, for instance the outside diameter of a pipe, differs from the target value. The closer the product characteristic is to the target value, the higher the quality.

The aim of process control is to prevent defects. On the other hand, the more traditional product control is aimed at detection of defects. There are subtle but very important differences between process and product control, see table 1-1.

The goal of product control is to make sure that the variability of a product characteristic is within the specification limits. This means that the product characteristic has to be below the upper specification limit, USL, and above the lower specification limit, LSL, see figure 1-1. The product characteristic in figure 1 is a length with USL=2.010", LSL=1.990", and a nominal target value of 2.000". Product A with a length of 2.000" will of course be accepted. Product B with a length of 2.009" will be accepted, however, product C with a length of 2.011" will not be accepted. In a detection based quality control system products A and B would both be considered good, while product C would be considered bad. It is clear, however, that products B and C are much closer together than products A and B. In fact, product A is much

TABLE 1-1. PRODUCT CONTROL COMPARED TO PROCESS CONTROL

	Product Control	Process Control
Focus	Product	Process
Goal	Variability within specification limits	On target with smallest variation
Tools	Acceptance sampling	Control charts
Improvement	Outgoing quality only	Quality plus productivity
Philosophy	Detection and containment of problems	Prevention of problems

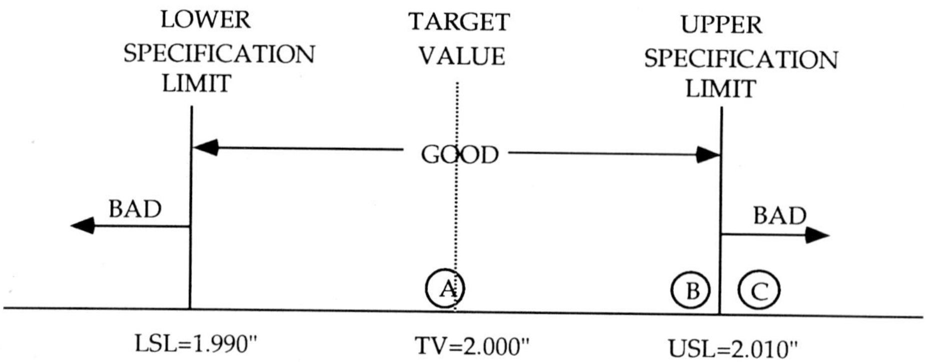

Figure 1-1 Concept of quality in detection based quality control

better than product B because A is right at the target value, while B is far from the target value, actually very close to the USL. In a prevention based quality control system product A would be considered best, product B fair, and product C bad, see figure 1-2.

The main tool in product control is acceptance sampling by inspection of finished goods. Some problems with this traditional approach to quality control are:

- inspection plan usually do not catch all the rejects
- when items are rejected the damage is already done
- inspection is time consuming and costly

A rejected product may be reworked to make it acceptable. However, there is additional cost associated with this rework. In the worst case, the product cannot be reworked and becomes an expensive piece of scrap.

Statistical process control involves quality control in each step of the production process. Standards are established for each step and an acceptable range about each

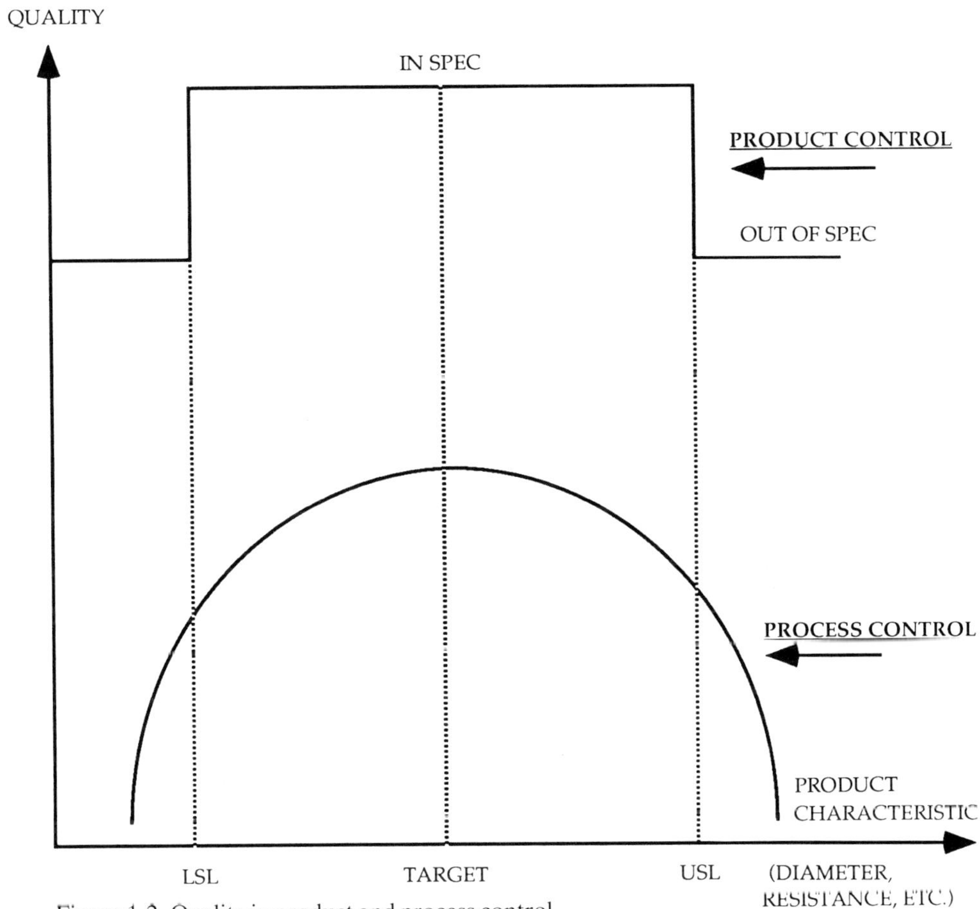

Figure 1-2 Quality in product and process control

standard is determined. As long as each process step yields a product within its set range, quality is ensured. When the acceptable range is exceeded, or a trend is identified which indicates the range will soon be exceeded, the process step is stopped and adjusted to bring it back in line with the standard.

It should be realized that SPC cannot solve all quality and production problems. It is not a cure-all! It will not correct a poor product design or poor employee training. Nor will it correct an inefficient process or worn out machines and tooling. However, it can help detect all of these types of problems and identify what corrective action is necessary to solve the problem.

One of the most important tools in SPC is the control chart. The most common control chart shows how the average of a set of measurements and the range of the measurements vary with time, see figure 1-3. The ***average*** of a set of measurements is often called the ***mean*** and represented by the symbol $\bar{x}$, pronounced x bar and sometimes written as x bar. The ***range*** is the difference between the largest and

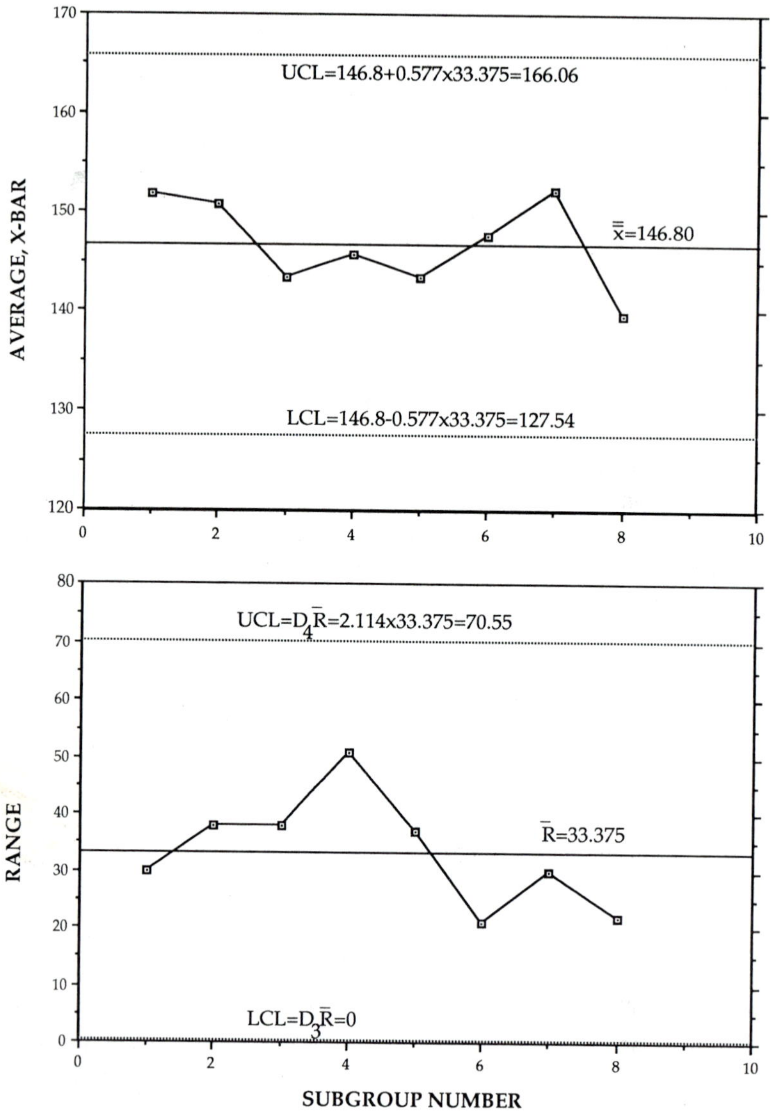

Figure 1-3 Example of a x-bar and R chart

smallest measurement, it is represented by the symbol R. This type of a control chart is called a $\bar{x}$ and R chart or the Shewhart control chart. Control charts can be used for controlling processes, for identifying problems, and for spotting trends before they become problems. Control charts will be covered in more detail in a later chapter.

The name of **Dr. Walter A. Shewhart**, mentioned above, is strongly linked to SPC. In fact, Dr. Shewhart can be considered the father of SPC. In the 1920s Dr. Shewhart developed methods for statistical monitoring of manufacturing processes. These methods with very little changes are today's SPC methods. Dr. Shewhart's book "Economic Control of Quality of Product", (1) published in 1931 was the first book to describe the basic elements of statistical process control.

Another name often linked to SPC is that of **Dr. W. Edwards Deming**, Dr. Deming, a former colleague of Dr. Shewhart, was instrumental in introducing SPC in Japan after the second world war. Ironically, even though SPC was developed in the U.S., it was largely ignored in the U.S. while Japan embraced the concepts of SPC under the tutelage of Dr. Deming. It is clear that one of the reasons for Japan's industrial success is the widespread acceptance and use of statistical process control.

As a result of increased competitive pressure SPC has made a resurgence in the U.S. in the 1980s. Some companies embrace SPC because they realize that it is a powerful tool to improve quality and productivity. Other companies adopt SPC because their customers require suppliers to use SPC and to supply SPC data with the product to prove that their production processes are in control and capable of making a good product consistently. An example is the American automobile industry; it has started to implement SPC not only in their own production facilities but also by demanding that their suppliers also implement SPC methods to ensure quality of subcontracted automobile components.

1.2 IMPLEMENTING STATISTICAL PROCESS CONTROL

One of the key issues in implementing SPC is training. Training has to occur at all levels: production workers, supervisors, and management. SPC gives production workers the tools to monitor their efforts and to improve the product being made, enhancing both pride in workmanship and job satisfaction.

Supervisors need to learn SPC so that they can make sure it is properly implemented on the work floor and so they can interpret SPC data in troubleshooting process problems. While SPC gives workers the proof that their work meets quality standards, it also gives both workers and supervisors the key to unlock the source of quality problems.

Management needs to learn SPC so that they can interpret SPC information in monitoring the output of their business. An added benefit is that statistical quality data can be a valuable product marketing tool. The SPC data gives management the proof of the quality of their product. As mentioned earlier, many companies require proof of quality from their suppliers of everything from raw material to semi-finished and finished goods. A properly documented SPC program provides all the proof of quality required to obtain those contracts.

The bottom line of SPC is profits. More profits arise from improved quality and productivity. Workers are given the tools to work smarter, not harder, to ensure quality with the goal of no rejects, no wasted effort. Supervision is given the means to coordinate production and quality control while eliminating the traditional antagonism between production and quality control functions. Management is given the tools to manage quality and reduce waste. The biggest selling point of SPC is more quality output for the same amount of input – increased productivity. Various tools are available to implement a statistical process control program. These tools include problem identification techniques, process control charts, and capability measures. Applying these tools, however, is no guarantee that the benefits of SPC will be attained. These tools are all used to find problems. The benefit of an SPC program is increasing quality and productivity through ***finding and eliminating*** problems.

One of the primary causes of failure of SPC is giving workers the tools to find problems without the means to solve these problems. Not being able to solve problems leads to inefficiency and frustration. Dr. Deming suggests that when a worker identifies a problem on the production floor, the person immediately responsible for that process is able to take corrective action and solve the problem only 6% of the time. The other 94% of the time the problem must be solved through ***cooperative action*** of various individuals: operators, supervisors, engineers, managers, etc. This means that effective management of SPC efforts is key to the successful implementation of SPC. The process of implementing SPC requires several steps to be taken.

The first step is the creation of an environment that allows and encourages problem solving. This step is often the most difficult one. SPC allows accurate detection of problems. However, detection of a problem is only part of the process of problem elimination. The next part of the process is recognition and acknowledgement of the problem by the appropriate people. This, invariably includes a person with supervisory responsibilities. In some cases, a supervisor or department manager may be reluctant to acknowledge a problem. This may be because he or she is afraid this creates a bad impression outside the department. The manager may then respond, not by acknowledging the problem, but by turning against the person who reported the problem. By turning against the bearer of bad news, the department manager has most likely made sure that nobody in the department will report a problem again. Consequently, this manager will be largely ignorant about problems in the department, making effective management next to impossible.

Creation of an atmosphere that allows problem solving, therefore, requires enlightened and capable supervisors and managers. This is an absolute necessity for efficient implementation of SPC. People should be encouraged to uncover and report problems. People reporting problems should be praised and rewarded, they should not be treated as trouble makers. The real trouble makers are the people ignoring problems, not the people reporting them!

The second step is training in SPC and process technology. In order to apply SPC successfully, the people involved with the process should have a basic familiarity with SPC concepts. As discussed earlier, training has to occur at all levels. However, training should cover not only statistical process control but also process technology. It is very difficult to apply SPC to extrusion if there is not a good understanding of the extrusion process. The same is true for other processes, such as injection molding, blow molding, compounding, etc. As a result, this book will not only cover basic SPC concepts but also basic extrusion technology.

The third step is to determine the key process problems and variables. Various methods can be used to identify and prioritize problems in SPC; these include: brain storming sessions, fishbone (cause and effect) diagrams, Pareto charts, histograms, scatter diagrams, etc. These methods will be discussed in detail in the next chapters. It is important to direct efforts to implement SPC to the most significant problems where an improvement in quality and productivity can be made and measured.

The fourth step is to use SPC to bring the process into control with its present set of conditions. These conditions can typically be grouped under the following main categories (the six m-words):

- man (personnel)
- machine
- material
- methods
- measurement
- milieu (environment)

In a later chapter it will be shown how process control charts can be used to reduce and eliminate causes of variation. The most commonly used control chart is the $\bar{x}$ and R chart shown earlier in figure 1-3.

The fifth step is to determine process capability. Once the process has been brought into control, it has to be determined whether the characteristics of the individual products meet the requirements of each characteristic. These requirements are often stated as engineering limits or product specifications. An example is an extruded sheet with a nominal thickness of 0.020" with an allowable upper limit of 0.021" and lower limit of 0.019". In this case the upper specification limit, USL, is 0.021" and the lower specification limit, LSL, is 0.019". Specification is often abbreviated as spec.

If the actual product variability is less than the specification width, the process to make the product is considered ***capable***, see figure 1-4. If the product variability is larger than the specification width, the process is considered not capable, see figure 1-5. In this case non-conforming products will be produced, this will result in rework or rejects. Various measures of process capability will be discussed in chapter 4.

If the process is not capable, some corrective action is necessary. **Thus, step six is to implement a plan of process improvement.** Various approaches are possible:

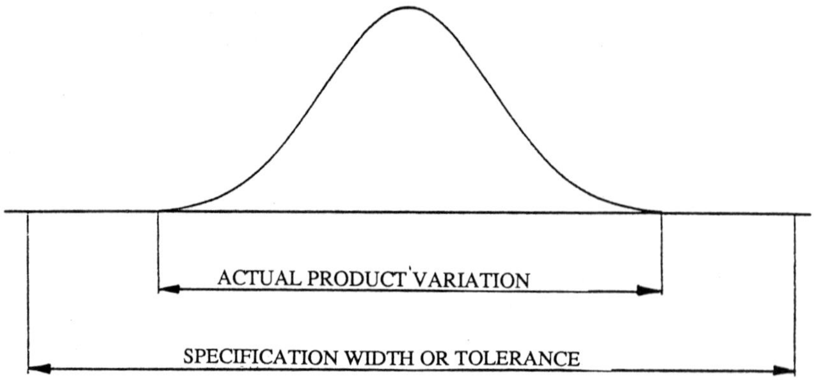

Figure 1-4 Example of a capable process

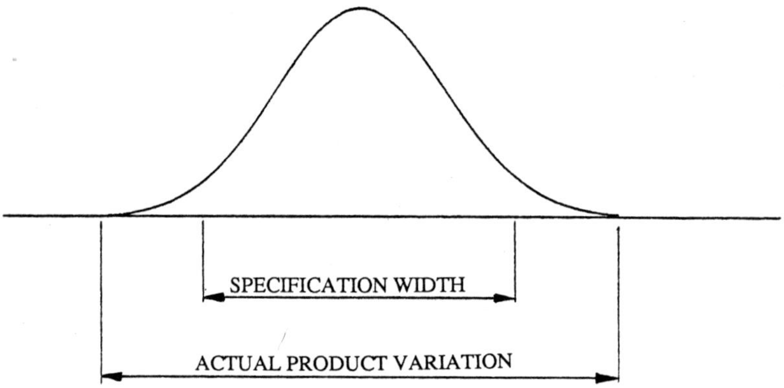

Figure 1-5 Example of a process that is not capable

- Improve process to reduce product variability.
- If the specification width is unnecessarily tight, the specification width could be broadened.
- Measure each product and eliminate out-of-specification products.
- Let the customer separate bad from good products.
- Quit making the product.

The last three approaches are usually not desirable. Broadening the specification width is often not possible. The most desirable corrective action, therefore, most often is to improve the process. This will typically involve a change in one of the conditions affecting the process. These are, as discussed earlier: man, machine, material, methods, measurement, and milieu; see figure 1-6. After a change has been made, we move back to step three and repeat the process until the process becomes ***capable***, see figure 1-7. In other words, until the product meet requirements consistently.

Next, we have to determine whether the products meet requirements for all important product characteristics, e.g. dimensions, color, surface finish, impact strength,

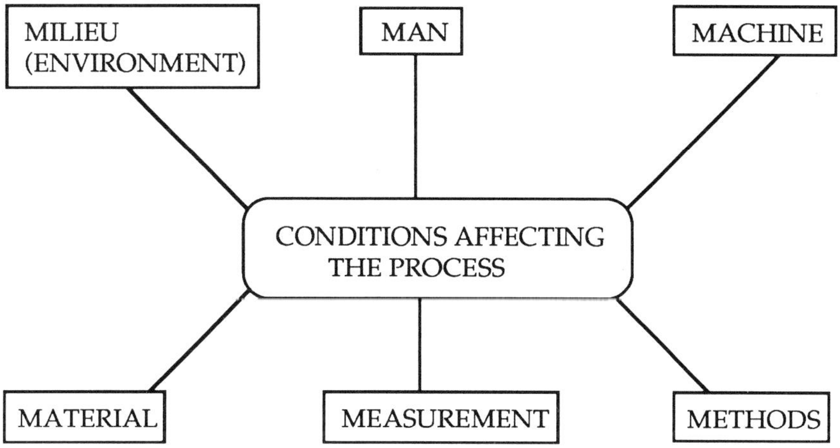

Figure 1-6 Main groups of conditions affecting the process

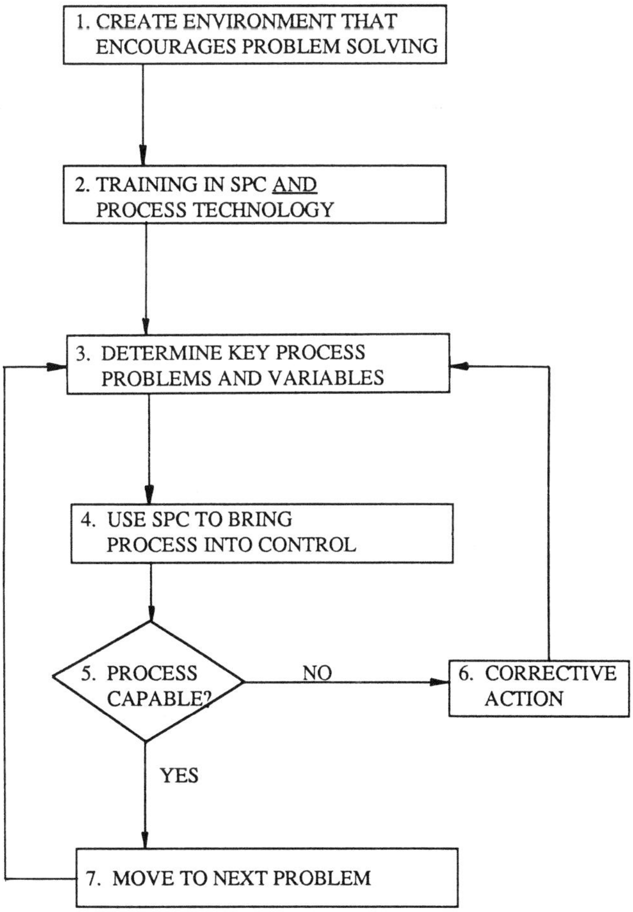

Figure 1-7 Implementation of statistical process control

tensile strength, etc. If the requirements for one characteristic are not met, then we have to take the next step. **The seventh step is to identify the next important problem.** We then move back to step three and repeat the various steps until all requirements can be met consistently, see figure 1-7. At this point, we can focus our attention to another process where problems may occur. If there is no other process, we can continue to work on the first process to further reduce variability.

1.3 BASIC STATISTICAL CONCEPTS

1.3.1 CAUSES OF VARIABILITY

All processes are subject to variability. Shewhart distinguished two basic causes of variability, ***common causes*** and ***assignable causes.*** A common cause is a source of variation that is inherent or natural to the process. They are also called ***random causes, chance causes, and natural causes.*** Examples of common causes in extrusion are ambient temperature, relative humidity, tooling adjustment, barrel temperature, etc. An assignable cause is a source of variation that is not inherent to the process but has an identifiable reason. They are also called ***special causes, sporadic causes, chaotic causes, and unnatural causes.*** Examples of assignable causes in extrusion are barrel heater burnout, unusual screw and barrel wear, contamination in raw material, etc.

The effect of a common cause is typically slight. No major part of the total variation can be traced to a single common cause. Common causes influence all data in a similar manner. They are stable and the collective pattern of common causes is predictable.

TABLE 1-2. CHARACTERISTICS OF COMMON CAUSES AND ASSIGNABLE CAUSES

	Common Cause	Assignable Cause
Nature	Inherent or natural to the process Predictable Stable	Unnatural Unpredictable Unstable
Typical Identity	Many small sources	One or few major sources
Effect	Slight	Can be strong
Improvement Action	Reduce common cause variability	Eliminate
Responsibility of Improvement Action	Management	Operator/Supervisor

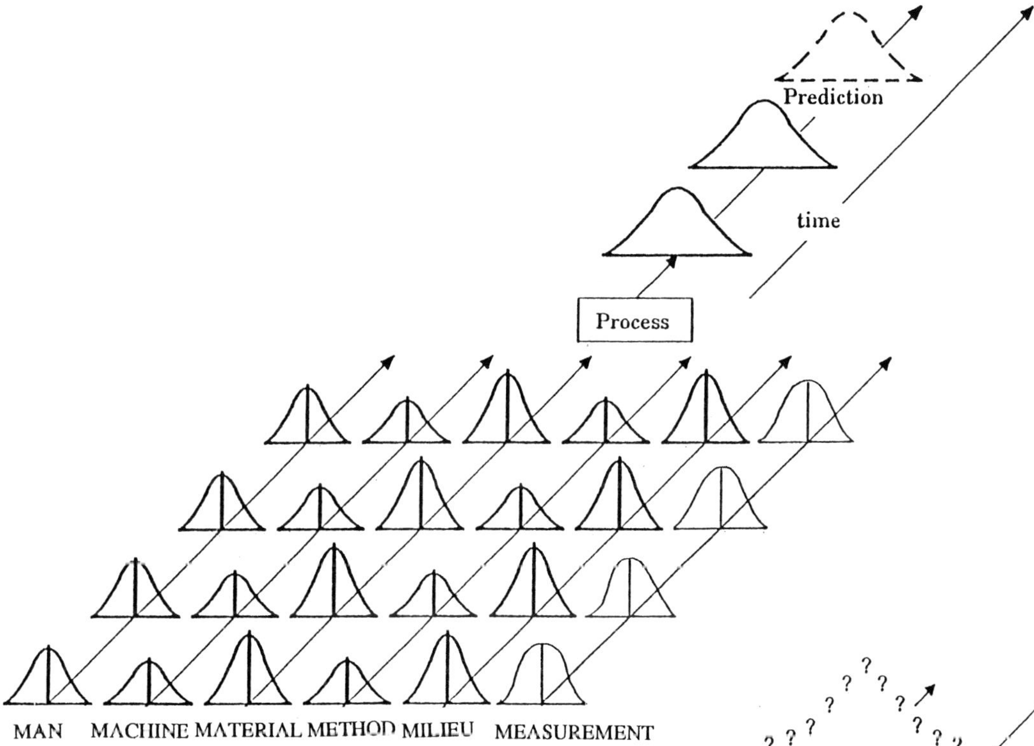

Figure 1-8 In-control process

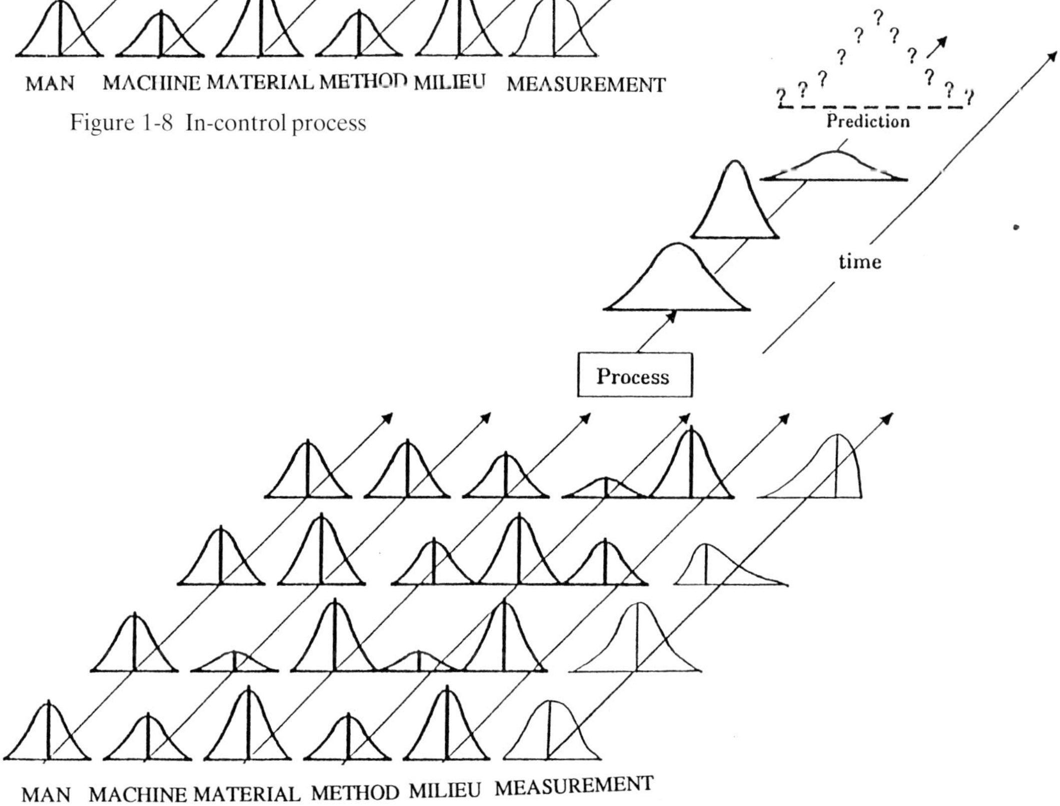

Figure 1-9 Out-of-control process

The effect of an assignable cause can be strong. Often a major part of the total variation can be attributed to a single assignable cause. Assignable causes influence some or all data in a dissimilar manner. They are not stable and the effect of assignable causes is unpredictable. Table 1-2 summarizes the characteristics of common and assignable causes.

If only common causes of variation are present, the output of a process forms a distribution pattern that is stable over time and predictable, see figure 1-8. If assignable causes of variation are present, the output of a process will not be stable over time and is not predictable, see figure 1-9. When we want to reduce process variability we usually focus first on variability due to assignable causes. There are a number of good reasons for this. Assignable causes often have a strong effect on total variability and they are usually easier to eliminate.

1.3.2 BASIC STATISTICAL TERMS

When we collect data on a process, we pull ***samples*** from the process. A sample is one or several individual pieces or measurements collected for analysis. Samples are usually pulled in small groups called ***subgroups***. Subgroups that are collected in a way that little variation can be expected between parts within the group, such as consecutive parts off an extrusion line, are considered ***rational subgroups.*** They are called rational because the way the subgroup has been chosen, hopefully, makes a certain amount of sense. Subgroups typically contain 5 to 10 individual pieces or measurements.

Subgroups, of course, are part of the total number of parts that are produced. All of the parts we manufacture make up a ***population*** or ***universe.*** By analyzing the variation of data in subgroups, we try to determine the variation in the population. A principal tool in the analysis is the control chart. Figure 1-10 shows the hierarchy of individual measurements, subgroups, and population.

1.3.3 MEAN, MEDIAN, AND MODE

In most processes we try to manufacture parts to a certain target dimension. This target dimension is called the nominal. Because of the natural variation in processes, the actual dimensions will not be exactly at nominal. Some will be higher and some will be lower. Hopefully, all dimensions will be close to the nominal. This tendency to group around a certain dimension is called ***central tendency.*** The most common measure of central tendency is the ***mean*** or ***average.*** It is the sum of the observations divided by the number of observations and is usually represented by the symbol $\bar{x}$. If we have a number of n observations with values x_1, x_2, x_3,.....x_n, the mean is calculated as follows:

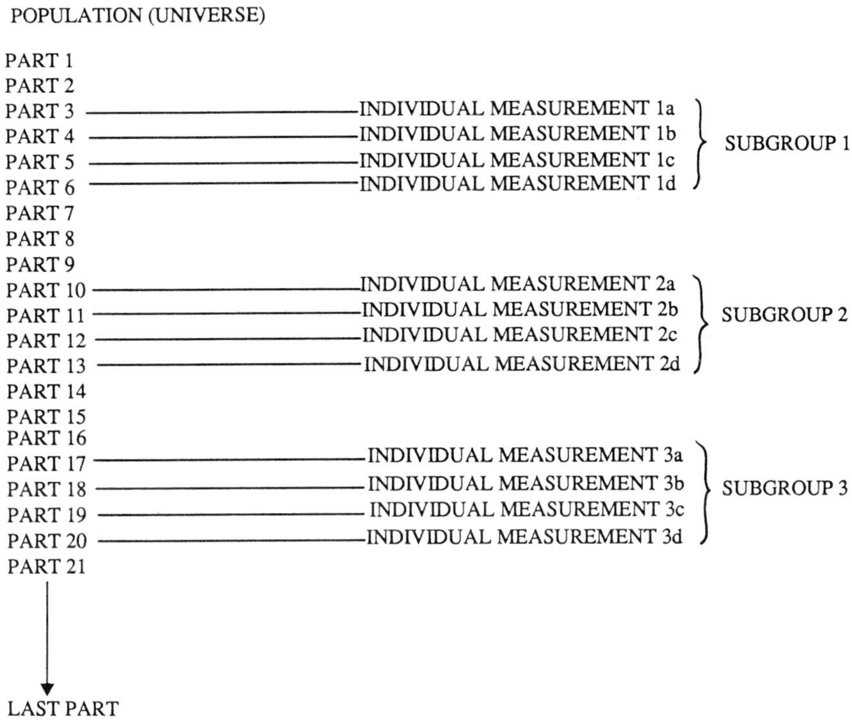

Figure 1-10 Relation between individual measurements, subgroups and population

$$\bar{x} = \frac{x_1 + x_2 + x_3 + \ldots\ldots x_n}{n} \qquad (1\text{-}1)$$

For example, we have 5 measurements of the thickness of extruded sheet:

$x_1 = 0.0203''$, $x_2 = 0.0208''$, $x_3 = 0.0197''$, $x_4 = 0.0198''$, *and* $x_5 = 0.0199''$

For this example the mean becomes:

$$\bar{x} = \frac{0.0203''+0.0208''+0.0197''+0.0198+0.0199''}{5} = 0.02010''$$

x̄ is often termed the first moment of the distribution; it is analogous to the center of gravity. Another measure of central tendency is the ***median***. It is the middle measurement with the numbers arranged in order of size. It is usually represented by x̃ (x-tilde), sometimes by the symbol Mi. For an even set of numbers, it is the average of the two middle numbers. If we arrange the numbers in the previous example in order of size we get:

0.0197"
0.0198"
0.0199" ← median
0.0203"
0.0208"

In this example the median is 0.199"

A third measure of central tendency is the ***mode***. It is the most frequently occurring value in a set of data. It can be represented by the symbol $\hat{x}$ (x-circumflex).

1.3.4 RANGE, VARIANCE, AND STANDARD DEVIATION

Mean, median, and mode are measures of central tendency. In addition to the central tendency, we also need to know about the spread of the data. Another term used for spread is dispersion. Commonly used measures for spread or dispersion are range, variance, and the standard deviation. The ***range*** is the difference between the largest and smallest measurement. The symbol for range is R; it is calculated from:

$$R = x_{max} - x_{min} \tag{1-2}$$

Variance is the mean of the squared deviations from the arithmetic mean. It is a kind of average that reflects the distance of individual measurements from the mean. The symbol for variance is v; it is calculated from:

$$v = \frac{(x_1 - \bar{x})^2 + (x_2 - \bar{x})^2 + (x_3 - \bar{x})^2 + ...(x_n - \bar{x})^2}{n-1} \tag{1-3}$$

where n is the number of measurements.
Variance is also called the second moment of the distribution. It is analogous to the moment of inertia. We will calculate the variance for the previous example with the measurements written as whole numbers:

$$x_1 = 203,\ x_2 = 208,\ x_3 = 197,\ x_4 = 198,\ and\ x_5 = 199$$

The mean was determined to be $\bar{x} = 201$. The variance now becomes:

$$v = \frac{(203-201)^2 + (208-201)^2 + (197-201)^2 + (198-201)^2 + (199-201)^2}{4} = 20.5$$

The ***standard deviation*** is simply the square root of the variance. The symbol for standard deviation is s; it is calculated as follows:

$$s = \sqrt{\frac{(x_1 - \bar{x})^2 + (x_2 - \bar{x})^2 + (x_3 - \bar{x})^2 + ...(x_n - \bar{x})^2}{n-1}} \tag{1-4}$$

Of course, we can also write the much shorter expression:

$$s = \sqrt{v} \tag{1-5}$$

Thus, the standard deviation for the example is

$$s = \sqrt{20.5} = 4.53$$

Another measure of dispersion that is used in some cases is the coefficient of variation or CV. The CV is the standard deviation divided by the mean:

$$CV = \frac{s}{\bar{x}} \tag{1-6}$$

The coefficient of variation is frequently used in the analysis of mixing. The inverse of CV is the signal to noise ratio; this is discussed in more detail under the Taguchi method in chapter 7.

The variance and standard deviation calculated so far are measures of the spread of the data in a sample. When we determine the variance and standard deviation of the entire population, we substitute the population mean μ for the sample mean $\bar{x}$. The population mean is usually represented by the Greek symbol μ (mu). The population standard deviation is typically represented by Greek symbol σ (sigma), while the sample standard deviation by symbol s.

If the population has a manageable size, such as 80 or 125, the population mean and standard deviation can be determined. However, when the population is very large, such as 450,000, it is often not practical to determine the mean and standard deviation of the population. In this case, the population mean is often approximated by the mean of the averages of several subgroups.

Variance has the property of additivity. This means that the total variance is equal to the sum of the variances of its parts. Standard deviation does not have the property of additivity. By analyzing the variances of different factors, it is possible to identify how these factors contribute to the total variance. This is know as the analysis of variances or ANOVA. It is an important tool in statistical process analysis. ANOVA will be covered in more detail in chapter 7.

The Greek characters are used to represent characteristics of the population, while Roman characters (regular alphabet) are used for characteristics of the samples. The standard deviation is a useful measure of the spread of data when the data follows a normal distribution. The normal distribution is very important in SPC; distribution patterns will be covered next.

1.3.5 PLOTTING DISTRIBUTION PATTERNS

All measurements are subject to variation. Their pattern can be graphically represented by distribution curves; these can be:

1. Histograms
2. Frequency polygons
3. Frequency curves

A histogram is basically a bar chart with the height of each bar indicating how many data occur within a certain interval; the width of the bar equals the width of the interval. To plot the distribution pattern, the following guidelines can be used:

A. Count the number of measurements (N)
B. Determine the number of intervals (K) from table 1–3
C. Determine the data range (R), this is highest minus lowest value
D. Estimate interval size by R/K
E. Round estimate interval to a convenient number
F. Determine interval boundaries
G. Determine one-half number accuracy
H. Adjust interval boundaries
I. Tabulate data

TABLE 1-3. GUIDE FOR NUMBER OF INTERVALS IN HISTOGRAMS

Number of Measurements (N)	Number of Intervals (K)
30 to 50	5 to 7
51 to 100	6 to 10
101 to 250	7 to 12
over 250	10 to 20

As an example we will examine measurements of wall thickness in extruded tubing expressed in inch:

0.0207	0.0198	0.0208	0.0199	0.0210	0.0197
0.0189	0.0192	0.0198	0.0197	0.0192	0.0191
0.0204	0.0201	0.0190	0.0203	0.0196	0.0194
0.0213	0.0208	0.0201	0.0207	0.0204	0.0206
0.0201	0.0197	0.0217	0.0201	0.0211	0.0204

To make the data easier to handle, we will write the data in whole numbers.

207	198	208	199	210	197
189	192	198	197	192	191
204	201	190	203	196	194
213	208	201	207	204	206
201	197	217	201	211	204

In this example the number of measurements N=30. From table 1-3 we select K=6. The highest value x_{max} = 217 and the lowest x_{min} = 189. Thus, the range R=217–189=28. The estimated interval size becomes 28/6=4.67. We will round the interval size to 5. We can now set the interval boundaries to 185, 190, 195, 200, 205, 210, 215, and 220. The one-half number accuracy is 0.5 and the adjusted interval boundaries become: 185.5, 190.5, 195.5, 200.5, 205.5, 215.5, and 220.5. When the data are whole numbers, the one-half number accuracy is always 0.5. The data can now be tabulated, see figure 1-11. The histogram as a proper bar chart is shown in figure 1-12. The frequency polygon plots the frequency in the center of the interval and connects the points by straight lines. This is shown in figure 1-13. By drawing a smooth line through the points, we obtain the frequency distribution curve, see figure 1-14.

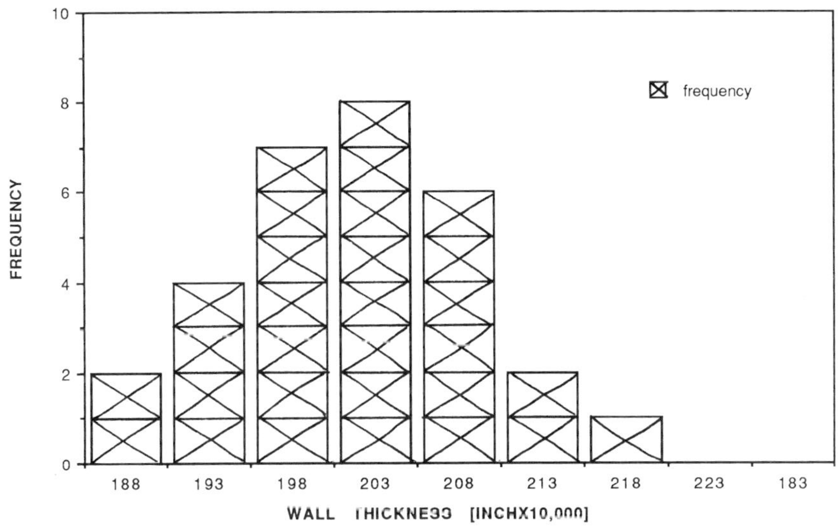

Figure 1-11 Wall thickness data tabulated by tallies

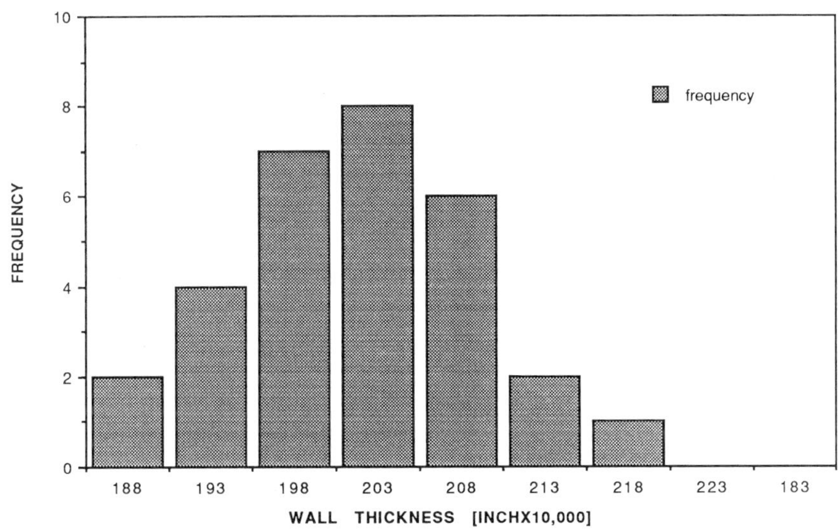

Figure 1-12 Bar chart of wall thickness data

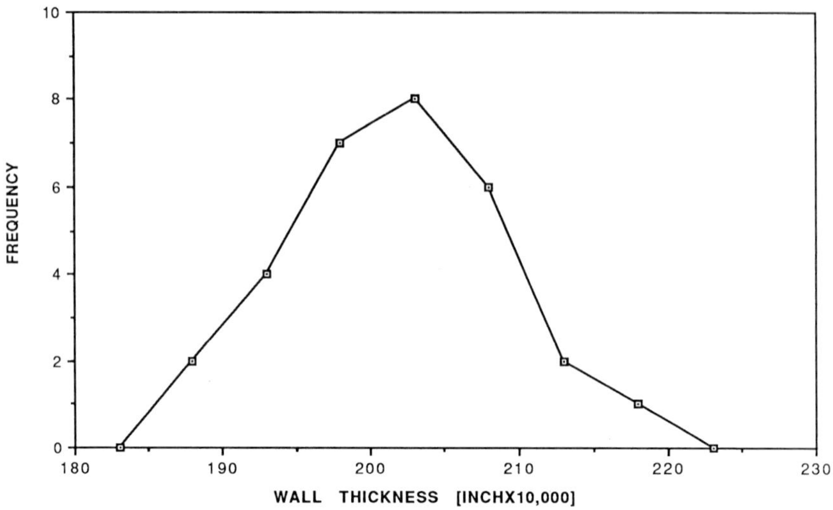

Figure 1-13 Wall thickness distribution shown as frequency polygon

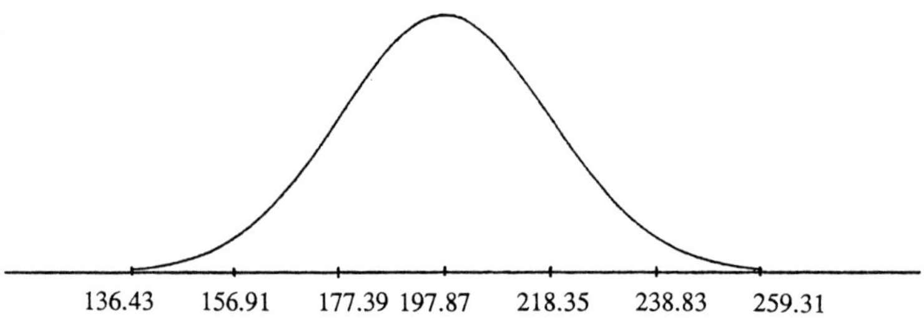

Figure 1-14 Smooth frequency distribution curve

1.3.6 CHARACTERISTICS OF A FREQUENCY DISTRIBUTION

Distributions can be described by their modality, variability, degree of skew, and degree of kurtosis. Some distributions have more than one point of concentration. These are said to be multimodal distributions. A distribution with a single point of concentration is called unimodal. Figure 1-15a shows a unimodal distribution, while figure 1-15b shows a bimodal distribution. If one distribution is narrow and another broad, both using the same horizontal scale, it means that the first has less variability than the second. This is illustrated in figures 1-15c and 1-15d.

A third characteristic is the symmetry of its variation; is it symmetrical or lopsided. If it is lopsided, the distribution is said to be skewed. Figure 1-15e shows a distribution with positive skew, while figure 1-15f shows a negatively skewed distribution. A distribution that is symmetrical has no skewness. The fourth characteristic deals with the relative concentration of data at the center and along the tails of the distribution. This characteristic is called kurtosis. If a distribution has a relatively high concentration in the middle and out on the tails, but relatively little in between, it has large kurtosis. If it is relatively flat in the middle and has thin tails, it has little kurtosis. This is illustrated in figure 1-15g and 1-15h.

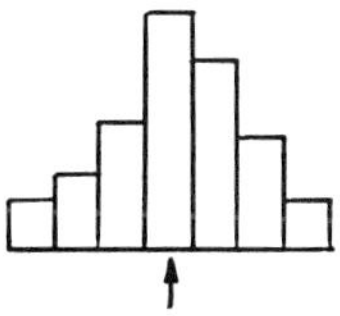

Figure 1-15a Unimodal

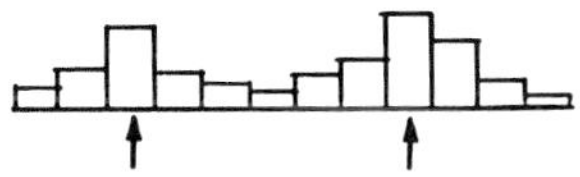

Figure 1-15b Bimodal

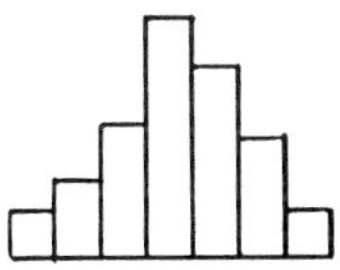

Figure 1-15c Small variability

Figure 1-15d Large variability

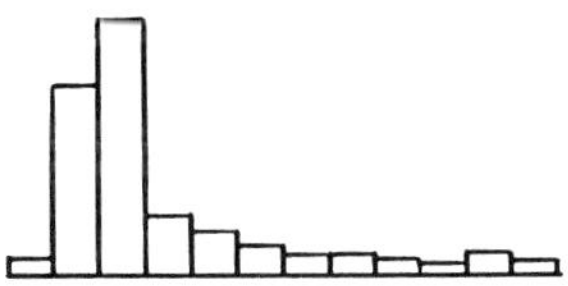

Figure 1-15e Positively skewed

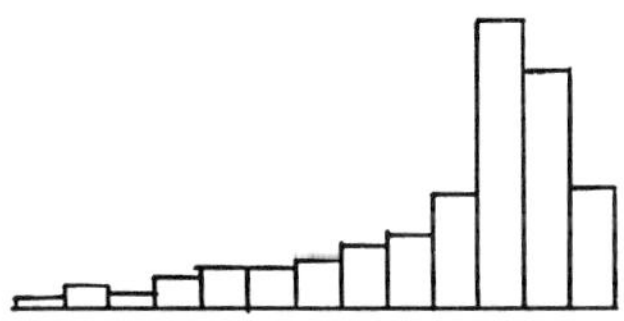

Figure 1-15f Negatively skewed

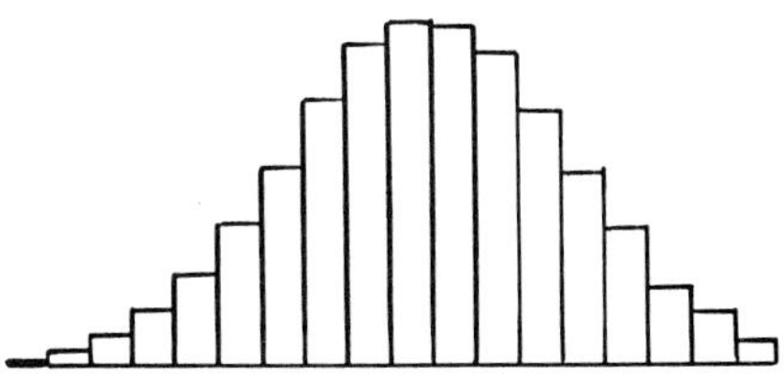

Figure 1-15g Normal kurtosis

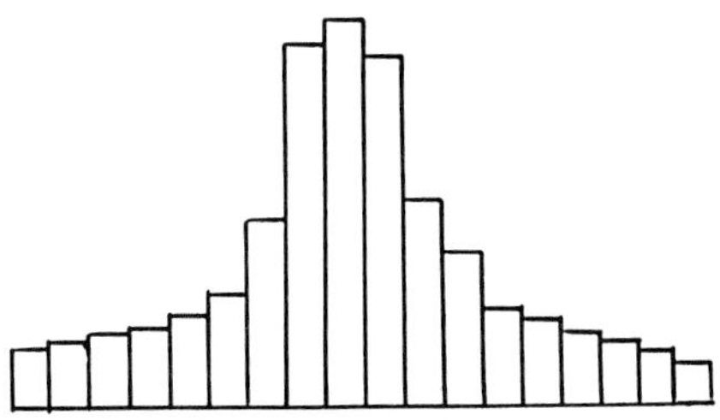

Figure 1-15h More than normal kurtosis

1.3.7 DIFFERENT DISTRIBUTION PATTERNS

Data can be distributed in a variety of different patterns. Some of the more important ones are:

1. The binomial distribution
2. The log normal distribution
3. The normal distribution

Other distributions are the Poisson and exponential distributions; these, however, are used less often in SPC.

The binomial distribution is used when the result of an inspection can only have one of two possible outcomes, such as good or bad. A product characteristic that can only be described by one of two possible conditions is called an ***attribute***. This is as opposed to a ***variable***, which is a product characteristic that can be described by one of many possible values. Attributes and variables will be covered in more detail later. The binomial distribution is useful in SPC when we deal with product attributes. When we are dealing with large samples, typically 50 or more, then the binomial distribution is very closely approximated by the normal distribution. This is true as long as the less probable of the two possible outcomes occurs at least 4 or 5 times in each sample.

The log normal distribution is a special type of normal distribution. Log is short for logarithm. We talk about a log normal distribution when a frequency curve of the logarithm of each of the individual points forms normal distribution curve. The log normal distribution is often used in the analysis of particle size distribution of solid materials, because many of them follow it closely.

The normal distribution is probably the most important one in SPC. It is also called Gaussian distribution. The distribution curve is bell shaped. It has the following characteristics:

- It is symmetrical about the mean, therefore the mean, median, and mode are all equal.
- It slopes downward on both sides to infinity.
- 68.25 % of all measurements lie between $\mu + 1\sigma$ and $\mu - 1\sigma$.
  See figure 1-16. The crosshatched area between $\mu + 1\sigma$ and $\mu - 1\sigma$ is 68.25% of the total area under the curve.
- 95.46% of all measurements lie between $\mu + 2\sigma$ and $\mu - 2\sigma$.
- 99.73% of all measurements lie between $\mu + 3\sigma$ and $\mu - 3\sigma$.

The equation for the bell-shaped normal distribution curve is presented in appendix IV. The area under a particular portion of the curve can be found in charts, usually called z charts. Such a z chart for standard normal distribution is shown in appendix IV. A graphical representation of the z chart is shown in figure 1–17.

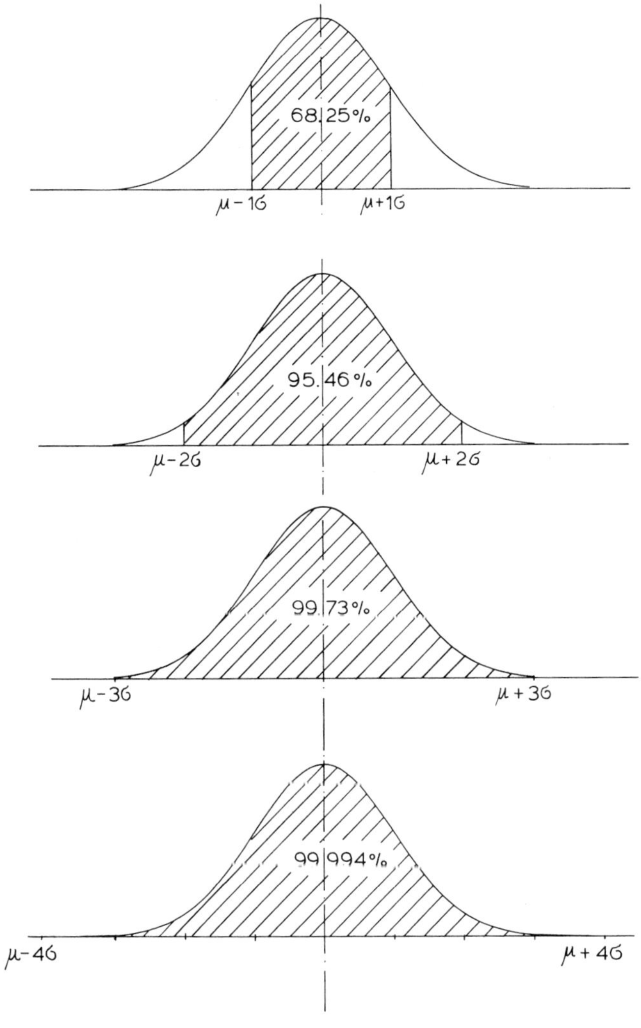

Figure 1-16 Areas under the curve for $\pm 1\sigma$, $\pm 2\sigma$, $\pm 3\sigma$, and $\pm 4\sigma$

1.3.8 THE CENTRAL LIMIT THEOREM

In using the normal distribution to determine how many measurements can be expected to be within certain values, we have to be sure that the data follow a normal distribution. The central limit theorem developed by Shewhart is very important in this respect. It states that sample averages (or x-bars) will follow a normal distribution as long as only common cause variations are present. This is true even if the individual measurements do not follow a normal distribution.

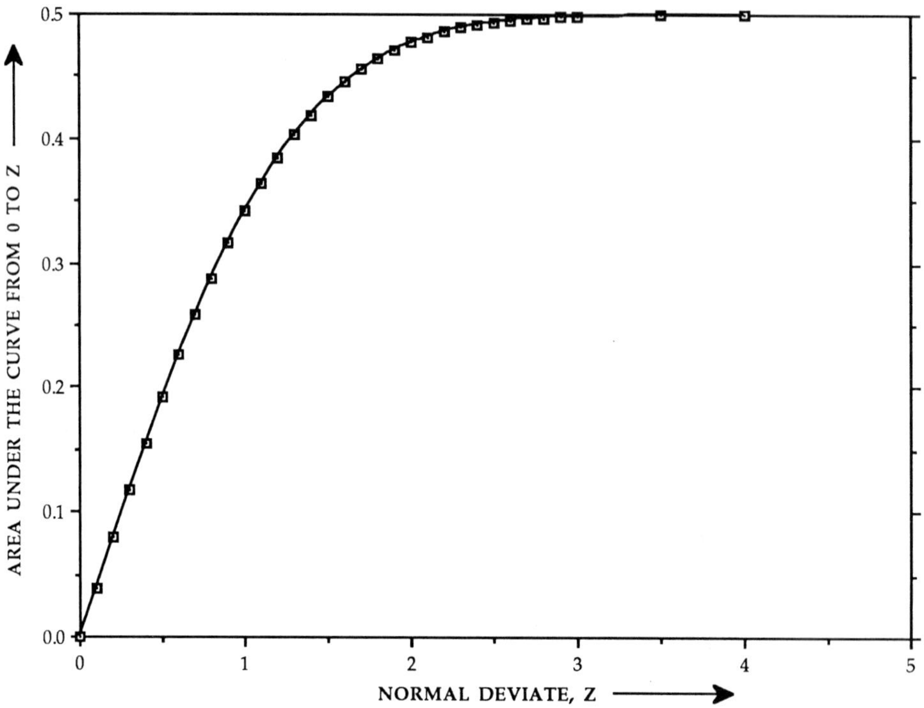

Figure 1-17 a Area under the normal curve from 0 to z

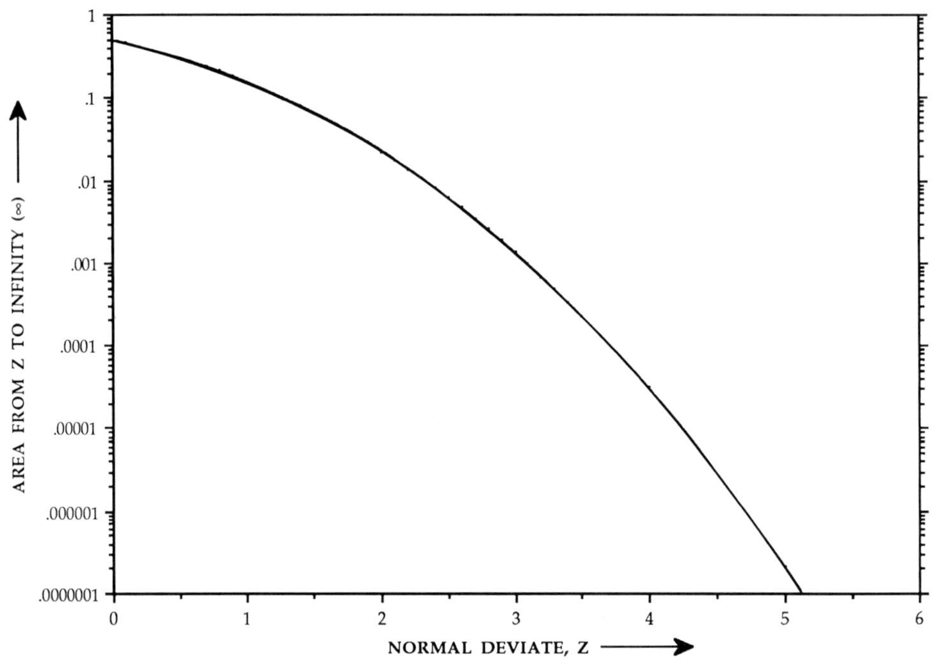

Figure 1-17 b Area under the normal curve on semilog plot

The standard deviation of the averages of subgroups, $\sigma_{\bar{x}}$, will be smaller than the standard deviation of the individual measurements, σ_x, see figure 1-18. The two are related by the formula:

$$\sigma_{\bar{x}} = \frac{\sigma_x}{\sqrt{n}} \tag{1-7}$$

where n is the number of data in the subgroup.

The charting of averages is the main element of control charts. The central limit theorem is the reason that control charts work.

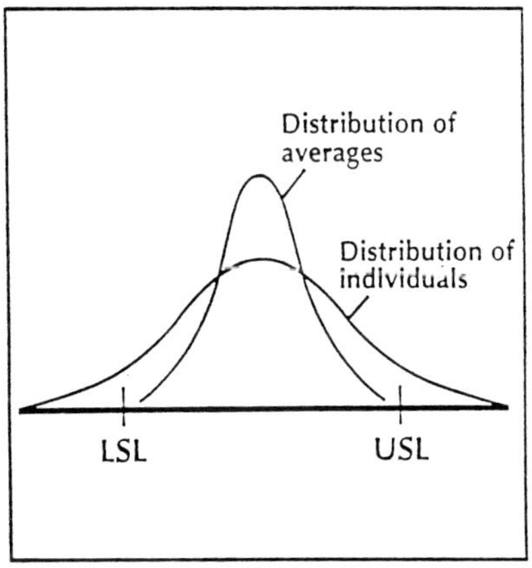

Figure 1-18 Comparison distribution of averages and distribution of individuals

2 DATA COLLECTION, DATA ANALYSIS, AND PROBLEM SOLVING

2.1 ATTRIBUTES DATA VERSUS VARIABLES DATA

As discussed in chapter 1, data can be divided into attributes data and variables data. Variables data is data that can be measured on a continuous scale. Attributes data is data that can have only one of two possible outcomes. A comparison between the two types of data is shown in table 2-1.

TABLE 2-1. COMPARISON ATTRIBUTES TO VARIABLES DATA

Variables Data	Attributes Data
Results from a measurement Example: 97.4°F, 16.24 oz.	Observations counted Example: yes-no, present-absent
Employs a measurement tool – Scale – Thermometer – Micrometer	Employs visual inspection – Go/No Go gauging – Pass/Fail inspection
Continuous scale	Discrete scale

When collecting data on the product, it is important to measure meaningful data. These are primarily determined by the functional requirements of the product and the customer requirements. In a continuous process, such as extrusion, it is most efficient to measure product characteristics ***on-line*** (the buzz-word that is used for this is real-time monitoring). This allows a rapid detection of problems and quick response before a large amount of out-of spec product is produced. Some product characteristics are easily measured on-line, such as outside diameter of extruded tubing with a laser gage. However, other characteristics may be difficult to measure on-line, such as wall thickness of small diameter extruded tubing. In such cases, the characteristic has to be measured off-line or measured indirectly. For example, in the example of wall thickness of tubing, the average wall thickness can be found by determining the output of the extruder and by measuring the line speed and tubing outside diameter. The output can be determined by using a gravimetric control or by using a gear pump with close to positive conveying characteristics. Measurements in the extrusion process will be discussed later in chapter 5.

2.2 IMPORTANT ASPECTS OF DATA COLLECTION

There are a number of important considerations related to data collection. They are measurement system evaluation, sampling methods, data handling, and process control systems. The analysis of a process is not meaningful unless the measuring instruments used to collect the data are both accurate and repeatable. Because of the importance of measurements, this subject will be covered in more detail in chapter 5 on measurement.

The accuracy of the analysis of a process is also dependent on the appropriateness of the sampling methods used. Data handling is an issue that can be important in data acquisition systems. These systems often sample at high rates and use some kind of averaging in the reporting of the data. This averaging, however, can mask certain process variations depending on the frequency of the variation and the averaging procedure. When using an electronic data acquisition system, EDAS, it is important to know how the data is sampled and analyzed internally. Process control systems will be discussed in chapter 6 on extrusion process technology.

2.3 DIAGRAMS FOR PROBLEM SOLVING

Various diagrams can be used to help in problem solving. ***Dr. Kaoru Ishikawa*** developed several such diagrams called cause and effect diagrams. As a result, they are often called Ishikawa diagrams. These diagrams are generally shaped as a fish skeleton; therefore, they are sometimes referred to as fishbone diagrams.
Three important cause and effect diagrams are:

- Cause listing diagram – a listing of the possible causes of a problem.
- Variation analysis diagram – a diagram used to analyze causes of variability in a process.
- Process analysis diagram – a flow diagram to study the various steps in a process.

These three diagrams look quite similar. The last two were derived from the first, the basic cause listing diagram.

2.3.1 CAUSE LISTING DIAGRAM

In a cause listing diagram the problem is put at the far right side of the page as shown in figure 2-1. This diagram lists a number of possible causes for voids in an extruded

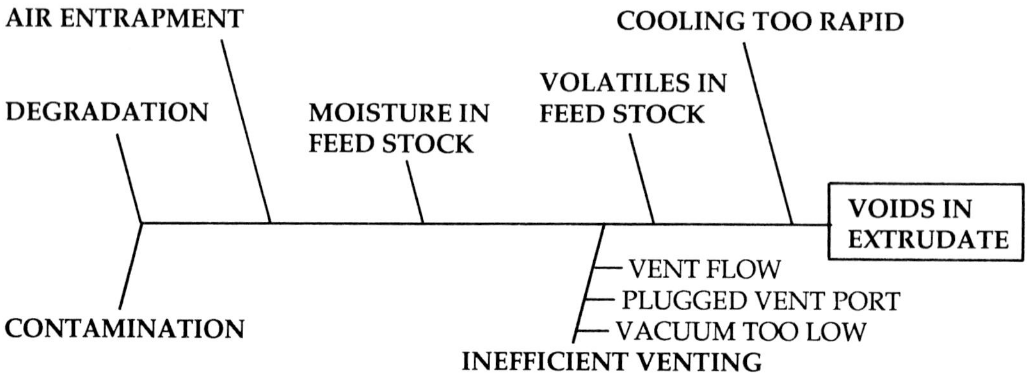

Figure 2-1 Cause listing diagram for voids in the extrudate

product. It is important to make this list as complete as possible. Therefore, it is important to solicit input from all the people associated with any aspect of the problem. A brainstorming session is a good way to come up with a large number of possible causes.

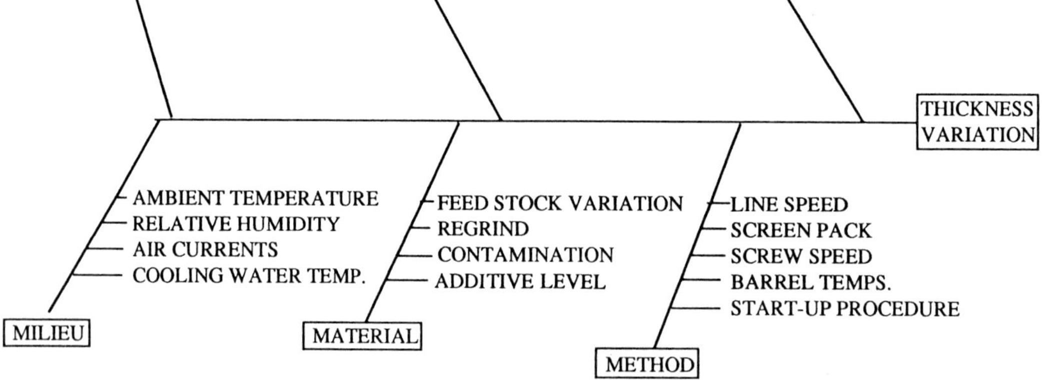

Figure 2-2 Possible sources of thickness variation in extrusion

2.3.2 VARIATION ANALYSIS DIAGRAM

A variation analysis diagram is constructed the same way as a cause listing diagram. The problem is listed on the far right. The main conditions affecting the process are listed: man, method, machine, material, measurement, and milieu. Next, all the details that can contribute to the variability are listed under the main categories. An example is shown in figure 2-2, which lists possible causes for thickness variation in an extruded product. The next step is to list details that can contribute to variability by adding arrows to the branches and labeling them. An example is shown in figure 2-3 for the main branch with material related sources of thickness variation. It is clear that several diagrams may be necessary to list all possible sources of variation.

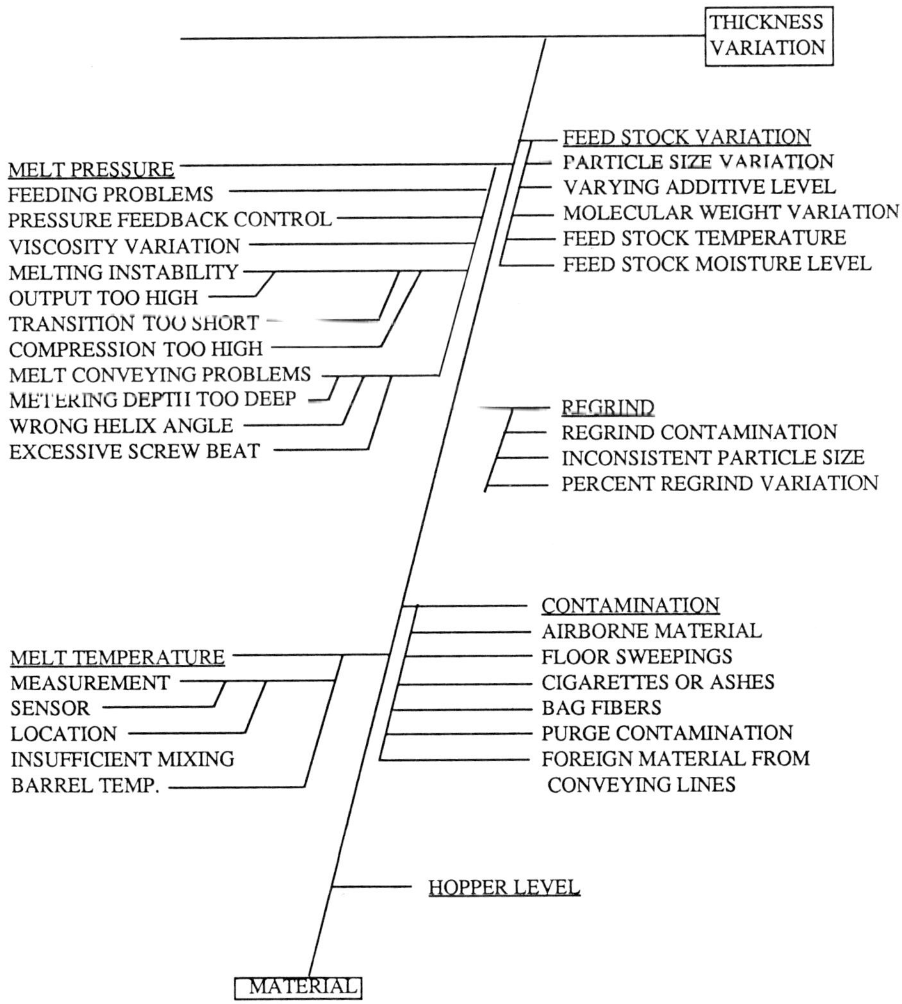

Figure 2-3 Material related sources of thickness variation

2.3.3 PROCESS ANALYSIS DIAGRAM

In a process analysis diagram each process step is labeled and connected by a line going from left to right, see figure 2-4. At each process step everything that can influence the product quality is added by branches. Figure 2-4 shows a process analysis diagram for a tubing operation with the problem being wall thickness variation. All the process steps are listed with factors that can potentially cause the problem.

The benefits of all these diagrams is that all critical factors have to be determined and put on paper. This allows a more systematic approach to problem solving. The drawback is, of course, that the diagrams are only as good as the people putting them together. It is important, therefore, to solicit input from as many sources as is reasonably possible. To summarize the benefits:

- All factors can be determined
- A systematic approach to problem solving can be taken
- The diagram provides a good record for future reference and may make problem solving easier and quicker in the future
- When the analysis has been completed, we can be reasonably sure that all factors have been accounted for.

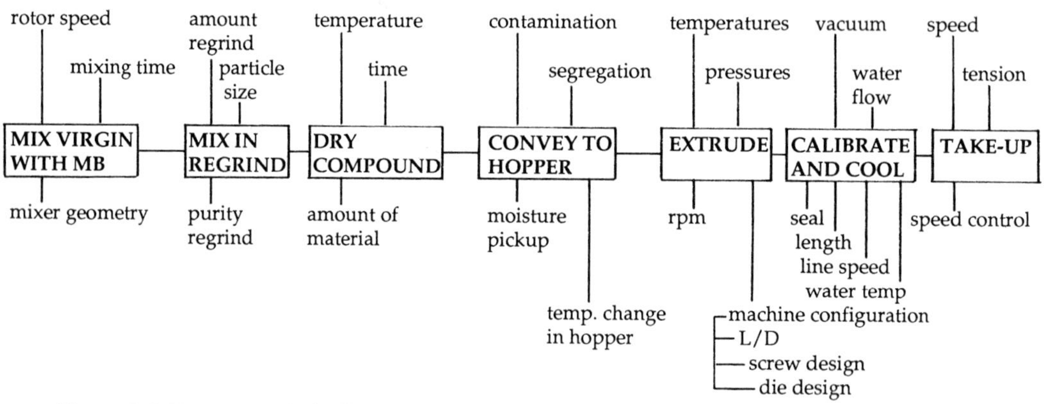

Figure 2-4 Process analysis diagram for simple tubing extrusion

2.4 PARETO DIAGRAMS

Pareto diagrams are a useful tool in separating the "critical few" from the "trivial many". In most processes the majority of the quality problems result from relatively few causes. Pareto diagrams show how many problems are caused by the various causes. The most frequently occurring problem is listed first, the second most frequent problem next, etc. The Pareto diagram is basically a bar chart with each problem represented by a different bar; the height of the bar represents the frequency of the problem.

Pareto diagrams can be used with both variables and attributes data, however, they are most often used with attributes data. The data is usually expressed in percentages or number of occurrences. We can take an example of defects in extruded film over a 24 hour time period as determined by visual inspection; the following data is obtained:

TABLE 2-2. TABLE SHOWING DEFECTS IN EXTRUDED FILM

Defect	Number	Percentage	Ranking
die lines	4	2.67%	5
black specks	7	4.67%	4
brown spots	78	52.00%	1
flow patterns	23	15.33%	2
fish eyes	12	8.00%	3
other	26	17.33%	
total	150	100.00%	

To construct the Pareto Diagram, it is convenient to rearrange the data as follows:

TABLE 2-3. TABLE SHOWING DEFECTS IN EXTRUDED FILM

Defect	Number	Percentage	Ranking
1. brown spots	78	52.00%	1
2. flow patterns	23	15.33%	2
3. fish eyes	12	8.00%	3
4. black specks	7	4.67%	4
5. die lines	4	2.67%	5
6. other	26	17.33%	
total	150	100.00%	

Figure 2-5 shows a Pareto chart showing the number of defects in each category on the left vertical axis and the percentage defects on the right vertical axis. Notice that the "other" category is shown last even though it may not have the lowest frequency. Also shown in figure 2-5 is a segmented line showing the cumulative percentage.

The main benefits of Pareto diagrams are:

- They uncover the few most important causes of problems.
- They can clearly demonstrate the results of improvements.

When using Pareto diagrams, it is important to use numbers of defects on the left-hand scale. This will clearly show the effect of improvement efforts; the percentages

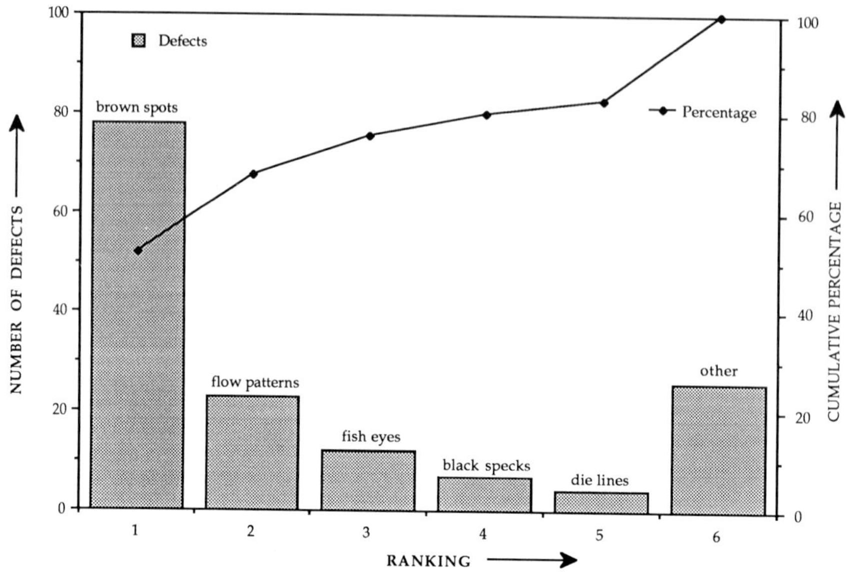

Figure 2-5 Pareto chart showing various defects in extruded films

will not clearly show the effect of improvements. Pareto diagrams, together with cause and effect diagrams, are useful in determining and directing the most efficient improvement efforts and in following the results of these efforts.

2.5 HISTOGRAMS

When data is collected on a process, we want to know how much variation occurs in the data. More specifically we want to know whether the data is centered around a single value, how wide the distribution is, whether the distribution resembles a normal distribution, etc. A histogram can answer questions. The construction and interpretation of histograms was discussed earlier in section 1.3.5.

2.6 SCATTER DIAGRAMS AND CORRELATION TABLES

When we have determined the causes contributing to a problem, we need to find out how strong an effect each cause has. One method of determining the cause and effect relationship is the scatter diagram, another is the correlation table. A scatter diagram iş simple a graph showing the cause on the horizontal scale and the effect on the vertical scale. Figure 2-6 shows the effect of extruder barrel zone temperature

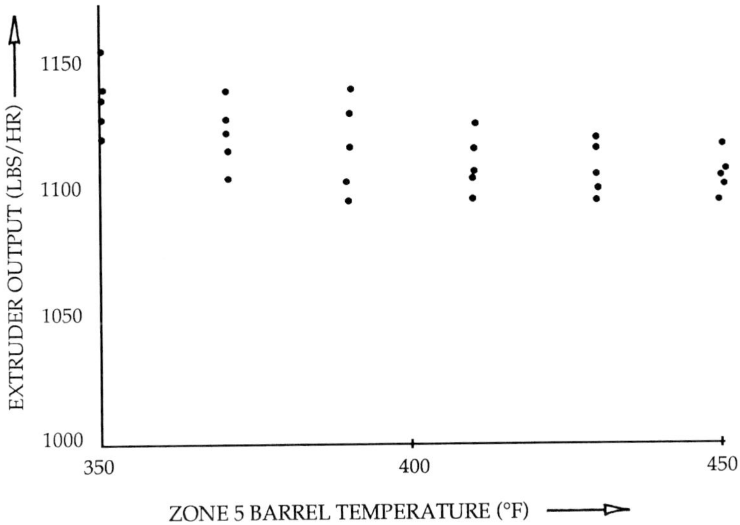

Figure 2-6 Scatter diagram relating extruder output to zone 5 temperature.

number 5 (the metering section) on extruder output. From figure 2-6 it is clear that increasing zone 5 temperature reduces the extruder output.

Another way of uncovering the cause and effect relationship is by using the correlation table. This table is useful with a large number of data and when many data have the same value. An example is shown in table 2-4. In this correlation table the effect of zone 1 barrel temperature on the pressure fluctuation at the end of the extruder screw is shown. The table shows that in this case the pressure fluctuation reduces with increasing zone 1 barrel temperature. Also listed in the table along the top and the right hand side is the number of data in each column and row.

TABLE 2-4. CORRELATION TABLE BARREL TEMPERATURE VS. PRESSURE FLUCTUATION

Number of Data→	10	10	10	10	10	50
200						0
180	*					1
160	**					2
140	*****	**				7
120	**	****	**			8
100		***	****	**	*	10
80		*	***	***	****	11
60			*	****	***	8
40				*	**	3
20						0
ΔP ≠	325	350	375	400	425	
(psi)	Barrel	Temp.	Zone 1	(°F)→		

When we plot a scatter diagram, a number of situations can occur. When the effect increases with the cause, we describe this as a positive correlation, see figure 2-7. When there is little scatter in the data, we will call this a strong positive correlation, see figure 2-7A. When there is substantial scatter in the data, this is called a weak positive correlation, see figure 2-7B. When the effect decreases with the cause, we have a negative correlation, see figure 2-8. Again, we can have a strong negative correlation, see figure 2-8A, or a weak negative correlation, see figure 2-8B. Finally, we can have a situation where there is no correlation at all, see figure 2-9.

It is possible to determine a value to describe the correlation between two factors. This is called a correlation coefficient (r). There are several methods of calculating correlation coefficients. In all methods r ranges from –1 to +1. A value of –1 indicates a perfect negative correlation, a value of +1 shows a perfect positive correlation, and a zero indicates no correlation. Correlation analysis will be discussed in more detail in chapter 7.

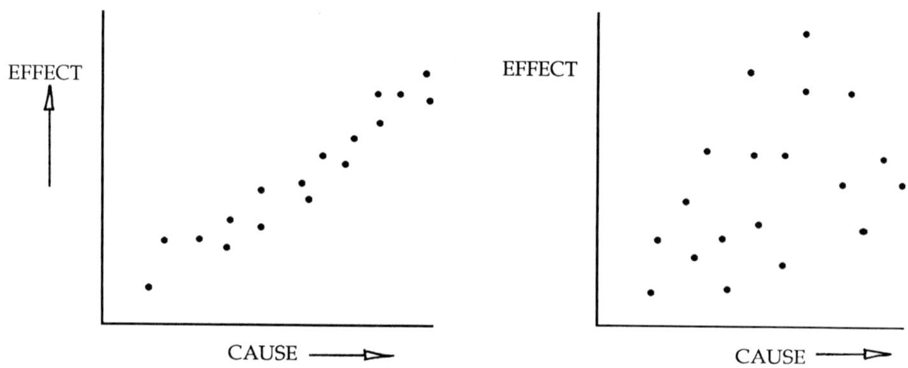

A, STRONG POSITIVE CORRELATION B. WEAK POSITIVE CORRELATION

Figure 2-7 Examples of strong and weak positive correlation

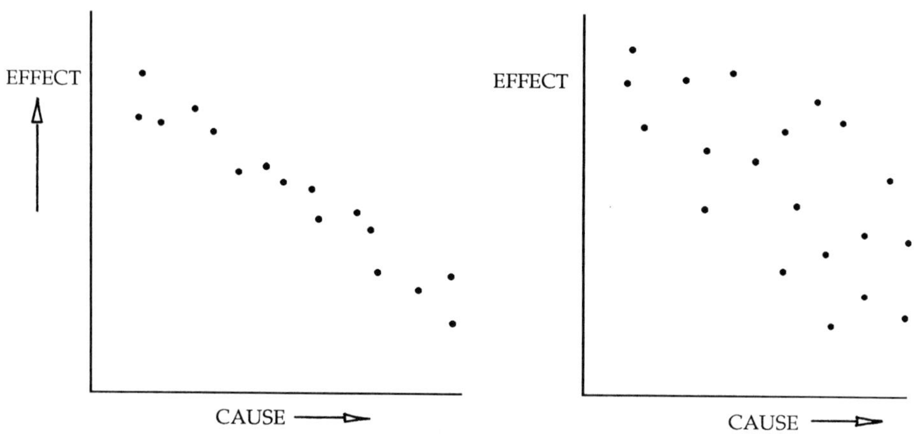

A, STRONG NEGATIVE CORRELATION B, WEAK NEGATIVE CORRELATION

Figure 2-8 Examples of strong and weak negative correlation

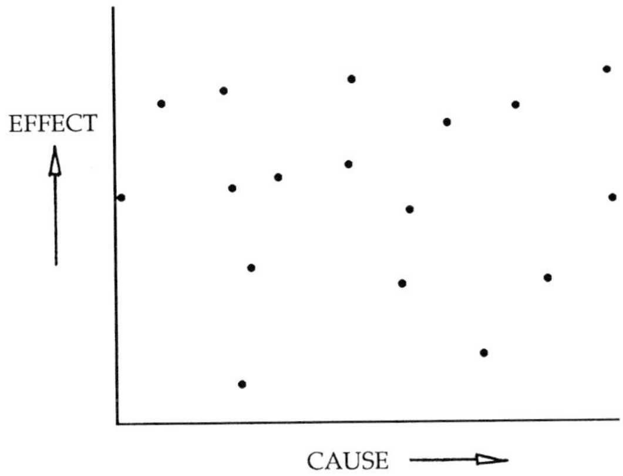

Figure 2-9 Example of no correlation

2.7 RECORDING DATA

There are three common ways to record data for process control. They are check sheets, portable data collectors, and fixed station data acquisition systems. Each method has advantages and disadvantages.

2.7.1 CHECK SHEETS

Check sheets are easy to make and the most flexible way to record data. They should be designed to allow efficient data collection with minimum chance of error. Errors occur more often when copying or reading data along a horizontal line than down a vertical column. Therefore, columns should be used for recording groups of data rather than rows. Pertinent information should be included in the check sheet such as the name of the data collector, the time and date, shift, machine, purpose of the study, etc. There should also be room for notes on the check sheet.

Check sheets can be used both for variables data and for attributes data. Variables data can be recorded in two ways, in columns with one column for each type of measurement or in a histogram when only one type of measurement is recorded on a check sheet.

A check sheet for variables with columns is shown in figure 2-10. This figure shows a check sheet to monitor cut tubing with three dimensions being measured; length, inside diameter, and outside diameter. A check sheet for variables in histogram form is shown in figure 2-11. The item being measured is the shrinkage of extruded sheet.

PRODUCT: TUBING
EXTRUSION LINE: NUMBER 3
MATERIAL: LDPE, LOT NUMBER 183
EXTRUDED: 3/27/91, WO. 326
SPECIFICATIONS: L = 4.000+/- 0.02O
ID = 0.180 +/- 0.010
OD = 0.300+/- 0.010
DATE/TIME: 3/29/91 7:20 AM
AUDITOR: M. NUNEZ
REMARKS:

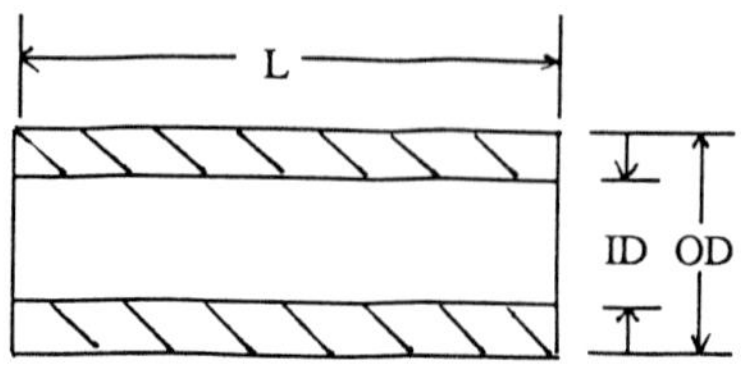

	L	OD	ID
	4.002	0.298	0.179
	4.007	0.299	0.183
	3.991	0.306	0.172
	3.998	0.291	0.185
	3.987	0.301	0.178
	4.012	0.303	0.184
AVERAGE	3.9995	0.2997	0.1802
RANGE	0.020	0.008	0.013

Figure 2-10 Check sheet for cut tubing

SHRINKAGE CHECK

PRODUCT: 5/8" PIPE
EXTRUSION LINE: A
MATERIAL: HDPE, COMPOUND 187
EXTRUDED: 3/14/91, WO2271

OVEN TEMPERATURE: 200 F
TIME IN OVEN: 1 HOUR
AUDITOR: M. SCHNERINGER
DATE/TIME: 3/16/91 9:40AM

REMARKS:

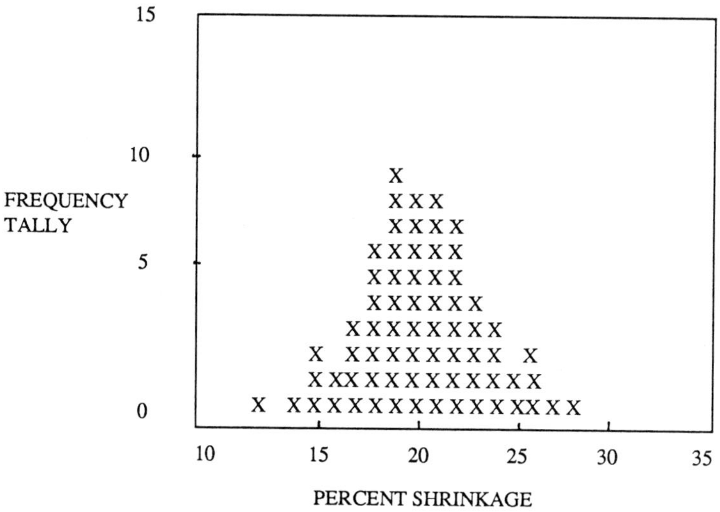

Figure 2-11 Check sheet for variables in histogram form

Check sheets for attributes data can be one of three types (see reference 2): defect-by-item, defect-by-location, and defect-by-cause. An example of a defect-by-item check sheet is shown in figure 2-12.

DEFECT TALLY SHEET

Date/time:	3-26-91/2:40 p.m.
Material:	4 mil 182 film for customer 2207, order XB701
Inspection:	F. Fernandez
Film Produced on:	Line 7, 3-17-91
Remarks:	

Defect	Number	Total
die lines	✓✓	2
black specks	✓✓✓✓✓	5
brown spots	✓✓✓✓✓ ✓✓✓✓✓ ✓✓✓✓✓ ✓✓✓✓✓ ✓✓✓✓✓ ✓✓✓✓✓ ✓✓✓✓✓ ✓✓✓✓✓	40
flow patterns	✓✓✓✓✓ ✓✓✓✓✓ ✓	11
fish eyes	✓✓✓✓✓ ✓	6
other	✓✓✓✓✓ ✓✓✓✓✓ ✓✓✓✓	14
	GRAND TOTAL	78

Figure 2-12 Defects-by-item check sheet for defects in extruded film

2.7.2 PORTABLE DATA COLLECTORS

Portable data collectors, PDCs, are similar to check sheets in that they can be easily moved around. However, they have a few important advantages:

- They can record data in computer readable form
- They can take data directly from electronic sensors and gages. This makes PDCs fast and minimizes errors.
- Data can be analyzed internally to yield information on mean, range, maximum value, minimum value, standard deviation, etc.
- Available with limit checking; the PDC can give an alarm when data just taken is out of specification.

PDCs can collect variables as well as attributes data. An example of a commercial PDC is shown in figure 2-13.

Figure 2-13 Example of a commercial PDC

2.7.3 FIXED STATION DATA ACQUISITION SYSTEMS

As the name implies, fixed station data acquisition systems are fixed to one location, either because of size or because of the wiring makes it very difficult to move the unit. A fixed station DAS can have a wide range of capabilities. A simple DAS may record data from only one source. A more sophisticated DAS may be able of recording data from various sources, analyze the data, present control charts, show trend plots for different variables, etc. Fixed station DASs can take many different forms depending on the application. DASs are available for extrusion systems. A good DAS can be a very valuable tool in improving process control and also in problem solving. For application to extrusion the following capabilities are useful:

- Monitoring of many variables. A typical extrusion process requires from 40 to 80 process parameters to be monitored.
- Data on each variable should be taken at least once a second. For some variables data should be collected at higher frequency, from 10 to 100 times per second. This is the case for variables where high frequency variation can affect extrudate quality. Examples are melt pressure, screw speed, line speed, extrudate dimensions, etc.
- Trending, the capability to display the variation of one or more process parameters over a particular time period. It is useful if the scales in these displays are adjustable. This capability is extremely helpful in trouble shooting and problem solving.
- Determination of statistical parameters on process and product parameters, such as mean, standard deviation, control limits, etc.
- Alarms for out-of-spec data and/or indications of assignable causes of variation.
- Recipes – the capability to store important process parameters for different products. This allows previous process conditions to be reproduced quickly and reliably.
- Production summaries – the capability to follow the amount of material being produced and present summaries per shift, per day, per week, etc. This is a useful management tool to analyze the productivity on the different production lines.

The following features have to be considered:
- *User friendliness*. The system should be easy to use, intuitive, and should not require extraordinarily long training for operators to become accustomed to the system.
- *Accessibility*. Is access from a remote computer possible? With remote access, integration into a plant wide information control system is possible.
- *Connectivity*. Can the DAS software work together with other software packages?
- *Upgradability*. Can new upgraded versions of software and/or hardware be readily implemented?
- *Cost*. An analysis should be made to determine whether the cost savings in terms of improved production efficiency and quality outweigh the cost of a data acquisition system.

The different methods of data collection each have their advantages and disadvantages. Check sheets are slow, prone to human errors and transcription errors, however, they are also inexpensive, flexible, efficient for small amounts of data, and easy to use. Fixed station data acquisition systems are fast, not prone to human errors and transcription errors, can handle large amounts of data, however, they tend to be expensive and more difficult to use. Thus, check sheets and fixed station data acquisition systems have complimentary areas of applications. Portable data collectors fall in between check sheets and fixed station data acquisition systems. Table 2-5 shows a comparison of the different data recording methods.

TABLE 2-5. COMPARISON DATA RECORDING METHODS

	Check Sheets	Portable Data Collectors	Fixed Station DAS
Speed	slow	fair	fast
Human errors	substantial	small	very small
Transcription errors	substantial	no	no
Efficient for	small amount of data	large amount of data	very large amount of data
Cost	minor	fair	substantial
Flexibility	very good	fair	good
User skill required	low	medium	high

2.8 SAMPLING

A population is the collection of all individuals in a designated group. Generally, populations are very large, making it impractical to collect data on each individual item. This is why we use samples. Sampling involves taking a small representative group from the population. For a sample to be representative, every item must have an equal chance of being selected for the sample. Sometimes random number tables are used so that the samples chosen are truly random. These samples are then evaluated for the characteristics that we are interested in, such as surface finish, length, diameter, and so on. The results from the samples are then translated back to estimate the characteristics of the population.

For most process control applications it is recommended that the sample size be kept small, as few as 3 or 4 items per sample and not more than 20. Rational subgroups or samples are collections of individual measurements or observations subject only to common cause variations. A system that is influenced only by common causes of variation is called a **common cause system**. Samples should be chosen to maximize the chance that the individual items within the sample are subject to a common cause system. Based on the following considerations, the sample size should be:

1). Small to minimize chances of special causes within the sample.
2). Large to ensure good characterization of the actual process.
3). Large to have the sample averages more closely follow a normal distribution
4). Large to provide sensitivity for the detection of process shifts.
5). Small to keep the data acquisition cost reasonable.

A sample size of 4 or 5 often provides a good balance of the various considerations. It is often tempting to combine the output of several processes that are assumed to be identical. For example, two equal size extruders side by side running the same product. However, such non-rational sampling can cause a number of problems, such as stratification and mixing. With stratification, as shown in figure 2-13, the x-bar control chart shows wide limits relative to the plotted points. This is due to the fact that the variability within the subgroup is subject to more than just common cause variation. It is important to monitor a single process only.

3 CONTROL CHARTS

3.1 INTRODUCTION

As discussed in chapter 1, control charts are one of the principal tools in statistical process control. Variations in a process can occur as a result of common causes and assignable causes. Common cause variation is inherent to any process that is stable over time. Assignable cause variation results from significant and identifiable changes in the process such as a different extruder screw, a change in the polymer drying conditions, a new extruder operator, a new lot of raw material, etc. In most cases reducing variation by eliminating assignable causes is more practical than by eliminating common causes. Process control charts are the best tool for identifying and eliminating assignable cause variation.

Control charts show how process data evolve over time and allow determination of natural process variability. Three important uses of control charts are:

- they can identify assignable causes when they occur.
- they can determine whether improvement action really reduces process variation.
- they can be used to determine the true process capability.

3.2 CONTROL CHARTS FOR VARIABLES DATA

Control charts can be used for both variables and attributes data. Variables control charts are more sensitive to changes in the measured values and, therefore, are better for process control. The various control charts for variables are shown in table 3-1.

TABLE 3-1. CONTROL CHARTS FOR VARIABLES DATA

Number of Data in Subgroup	Type of Control Charts
$n = 1$	x & Rm
$1 < n < 7$	$\bar{x}$ & R, $\tilde{x}$ & R, $\bar{x}$ & s, CuSum
$n > 7$	$\bar{x}$ & s, CuSum

x & Rm – The individual measurement, x, and moving range, Rm, chart is used when process control is based on individual readings. This can happen when the measurements are expensive. Sometimes the moving range chart is combined with a moving average chart ($\bar{x}m$ & Rm).

$\bar{x}$ & R – The mean, $\bar{x}$, and range, R, chart is the most common control chart. It is discussed below in more detail.

$\tilde{x}$ & R – The median, $\tilde{x}$, and range, R, chart is similar to the $\bar{x}$ & R chart. The median is the middle value when data is arranged according to size. Sometimes the median chart is used by itself without the range chart.

$\bar{x}$ & s – The mean, $\bar{x}$, and sigma, s, chart is a more accurate indication of process variability than the $\bar{x}$ & R chart, particularly with a larger size subgroup. The standard deviation, s, is more difficult to calculate that the range, R, however, this is immaterial when the data is processed by a computer.

CuSum – A cumulative sum (CuSum) chart shows the cumulative sum of each $\bar{x}$ value minus the nominal value. This chart is more sensitive to a sustained shift from the nominal.

3.2.1 THE x-BAR AND R CHART

Since the $\bar{x}$ & R chart is so commonly used in SPC, we will treat this chart in depth. The methods of usage and analysis for other charts are similar to the $\bar{x}$ & R chart. Therefore, what we learn about the analysis and interpretation of $\bar{x}$ & R charts can be applied to other charts as well.

The creation of an $\bar{x}$ & R chart requires several steps. We will go through this process step by step.

Step 1. Select the subgroup size and how often samples are taken. A common subgroup size is n = 5, with n the number of measurements in the subgroup.

Step 2. Record the raw data on the control chart. The data in each subgroup is listed columns. If the number of subgroups is k, then the total number of measurements is n x k. Proper determination of control limits requires at least 25 subgroups (k≥25)

Step 3. Calculate the sum, Σx, the mean, $\bar{x}$, and the range, R, for each subgroup using the following expressions:

$$\sum_{i=1}^{i=n} x_i = x_1 + x_2 + x_3 + \ldots\ldots\ldots\ldots + x_n$$

$$\bar{x} = \Sigma x_i / n$$

$$R = x_{max} - x_{min}$$

Σx_i is a simplified notation for $\sum_{i=1}^{i=n} x_i$, which is simply a shorthand notation for the sum of the x-values from x_1 all the way through x_n

Step 4. Calculate the central lines for the x-bar and R chart.

$$\bar{\bar{x}} = (\bar{x}_1 + \bar{x}_2 + \bar{x}_3 + \dots\dots\dots\dots\dots + \bar{x}_k) / k$$

$$\bar{R} = (R_1 + R_2 + R_3 + \dots\dots\dots\dots\dots + R_k) / k$$

where:

$\bar{R}$ = the average range

$\bar{\bar{x}}$ = the average of all samples or the "grand average".

Step 5. Calculate the control limits for the x-bar and R chart.

$$\text{UCL x-bar} = \bar{\bar{x}} + A_2\bar{R}$$

$$\text{LCL x-bar} = \bar{\bar{x}} - A_2\bar{R}$$

$$\text{UCL range} = D_4\bar{R}$$

$$\text{LCL range} = D_3\bar{R}$$

UCL is the upper control limit and LCL is the lower control limit. The constants A_2,D_4, and D_3 can be taken from table 3-2.

TABLE 3-2. FACTORS FOR DETERMINATION OF CONTROL LIMITS

Subgroup Size n	Factor A_2	Constant d_2	Factor D_3	Factor D_4
2	1.880	1.128	0	3.267
3	1.023	1.693	0	2.574
4	0.729	2.059	0	2.282
5	0.577	2.326	0	2.114
6	0.483	2.534	0	2.004
7	0.419	2.704	0.076	1.924
8	0.373	2.847	0.136	1.864
9	0.337	2.970	0.184	1.816
10	0.308	3.078	0.223	1.777
11	0.285	3.173	0.256	1.744
12	0.266	3.258	0.283	1.717
13	0.249	3.336	0.307	1.693
14	0.235	3.407	0.328	1.672
15	0.223	3.472	0.347	1.653
16	0.212	3.532	0.363	1.637
17	0.203	3.588	0.378	1.622
18	0.194	3.640	0.391	1.608
19	0.187	3.689	0.403	1.597
20	0.180	3.735	0.415	1.585

Rauwendaal Extrusion Engineering, Inc. $\bar{x}$ & R chart for n=5

Product name/number: Dimension charted:								Starting date/time: Ending date/time:				
sample	1											
data	2											
	3											
	4											
	5											
sum												
mean, $\bar{x}$												
range, R												

UCL ($\bar{x}$) = $\bar{\bar{x}} + 0.577\bar{R}$ =   UCL (R) = $2.114\bar{R}$ =
LCL ($\bar{x}$) = $\bar{\bar{x}} - 0.577\bar{R}$ =   LCL (R) = 0

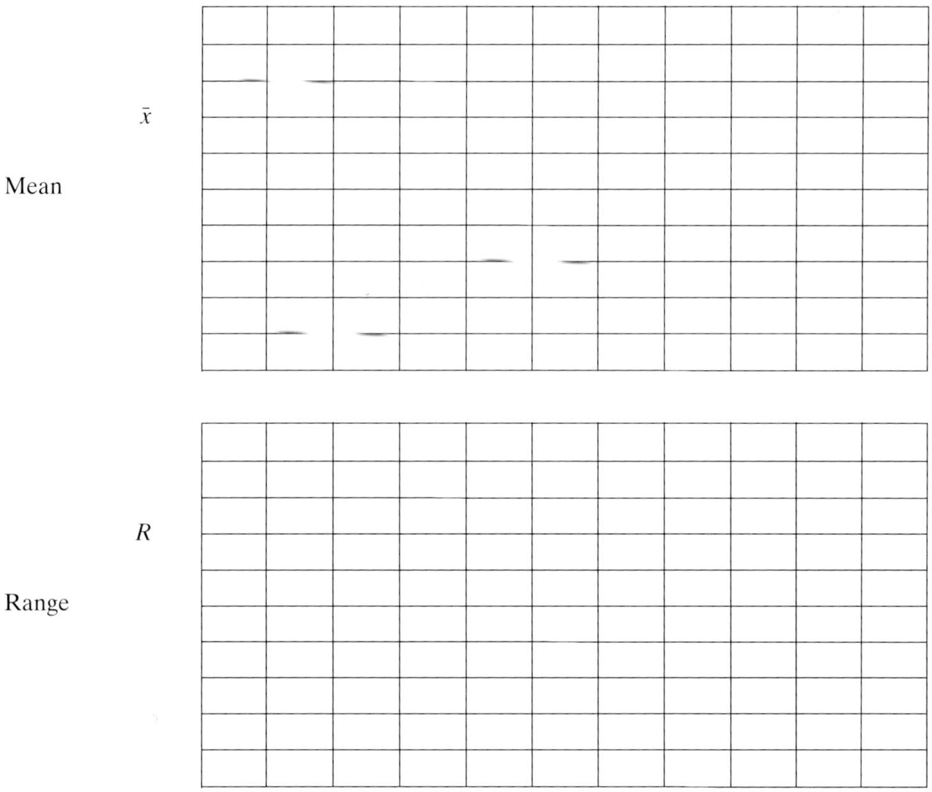

Figure 3-1 Blank form for x-bar and R charts

Step 6. Plot the sample averages and the sample ranges. Use the following guidelines:

– The grand average, $\bar{\bar{x}}$, should be near the center of the $\bar{x}$ chart.

- The UCL for the x-bar chart should be at about 2/3 from the $\bar{\bar{x}}$ to the top of the scale.
- The UCL and LCL have to be equidistant from $\bar{\bar{x}}$.
- The scale of the R chart should extend from 0 to about 4/3 times the UCL range.
- The points should be plotted as solid dots.
- Connect the points for easier pattern identification

Step 7. Draw the central lines and control limits on the charts. The central lines should be solid, the control limits dashed. The lines should be labeled with their numerical value.

The data can be tabulated and plotted on regular paper or special chart forms can be used to make organization of the data easier. A blank form for an $\bar{x}$ & R chart for subgroups of 5 is shown in figure 3-1. An example of a tabulated data of the thickness of extruded sheet is shown in figure 3-2. For this example: $\bar{\bar{x}} = 146.8$, $\bar{R} = 33.375$, UCL ($\bar{x}$) = 166.06, LCL ($\bar{x}$) = 127.54, UCL (R) = 70.55, and LCL (R) = 0. The $\bar{x}$ & R charts are shown in figure 3-3. Now that we know how to construct $\bar{x}$ & R charts, the next important step is the analysis and interpretation of $\bar{x}$ & R charts.

LOWEST VALUE		138	150	164	132	119	144	144	140
		146	158	140	125	147	149	163	135
		161	138	126	147	153	157	154	150
HIGHEST VALUE		168	173	142	148	142	136	165	145
		146	135	145	176	156	152	135	128
AVERAGE	$\bar{X}$ =	151.8	150.8	143.4	145.6	143.4	147.6	152.2	139.6
RANGE	R =	30	38	38	51	37	21	30	22

GRAND AVERAGE $\bar{\bar{X}} = (151.8 + 150.8 + 143.3 + 145.6 + 143.4 + 147.6 + 152.2 + 139.6)/8 = 146.8$

AVERAGE RANGE $\bar{R} = (30 + 38 + 38 + 51 + 37 + 21 + 30 + 22)/8 = 33.375$

Figure 3-2 Tabulated thickness data

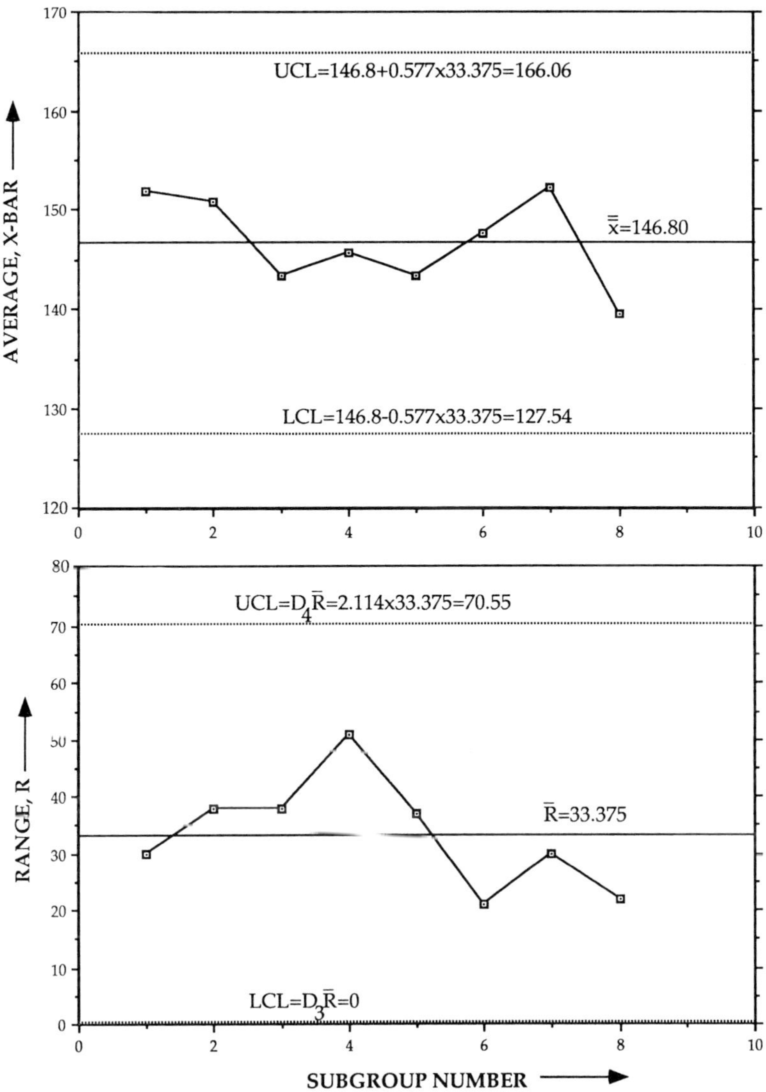

Figure 3-3 Example of x-bar and R control chart

3.2.2 INTERPRETATION OF x-BAR AND R CHARTS

The most common feature of a process showing stability is the absence of any recognizable pattern, see figure 3-4. The three main characteristics of a stable process are:

- Most points are near the center line
- Some points are spread out and approach the control limits
- No points beyond the control limits.

CHARACTERISTICS OF A STABLE PROCESS

* ABSENCE OF ANY RECOGNIZABLE PATTERN

* MOST POINTS NEAR THE CENTERLINE

* SOME POINTS SPREAD OUT AND APPROACH CONTROL LIMIT

* NO POINTS BEYOND THE CONTROL LIMITS

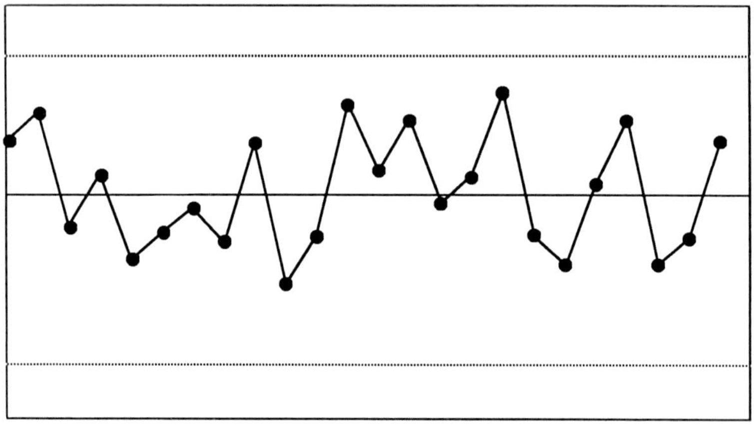

NORMAL PATTERN

Figure 3-4 Example of control chart of stable process

Characteristics of an unstable process influenced by assignable causes are:

- Freaks. One or more points outside the control limits, see figure 3-5.
- Trend. Six consecutive increases or decreases, see figure 3-6.
- Cycle. A repetitive pattern, see figure 3-7.
- Run. Seven or more points in a row above or below the center line, see figure 3-8.
- Zone rules:
  1. Two of three successive points fall beyond 2 standard deviations from the mean.
  2. Four of five successive points fall beyond one standard deviation from the mean, see figure 3-9.
- Shift in level. A sudden change in the central tendency of the data, see figure 3-10.
- Stratification. 15 or more points very close to the center line, see figure 3-11.
- Clusters. The grouping of points in one area of the chart, see figure 3-12.
- Mixture. Identified by an absence of points near the center line, see figure 3-13.

To identify assignable causes it is best to focus first on the R chart. The R chart is more sensitive to changes in uniformity and consistency. Any change in the process will tend to shift the R values upward. Since R charts are more sensitive to changes in the process, the effect of process improvement will be noticed first in the R chart.

The $\bar{x}$ chart should be analyzed after the R chart is stable. The $\bar{x}$ chart can be misleading when the R chart is unstable. When both the $\bar{x}$ and R charts are stable, the

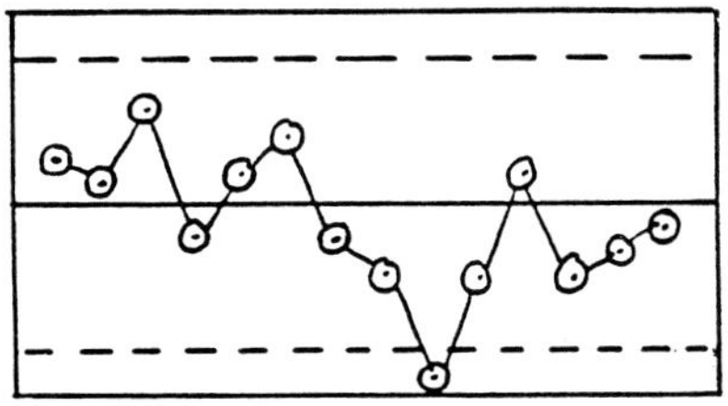

Figure 3-5 Point outside control limits

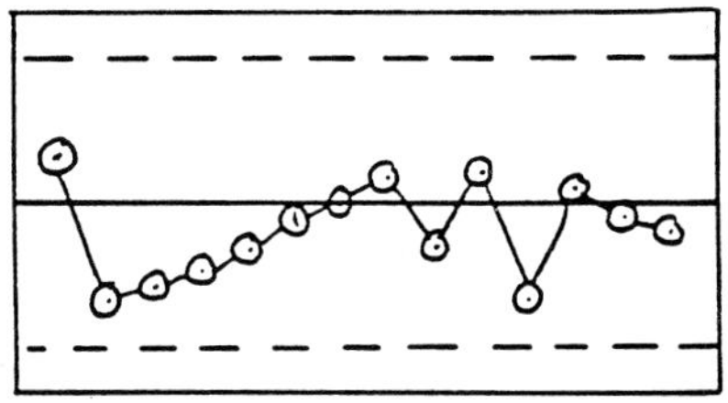

Figure 3-6 Trend

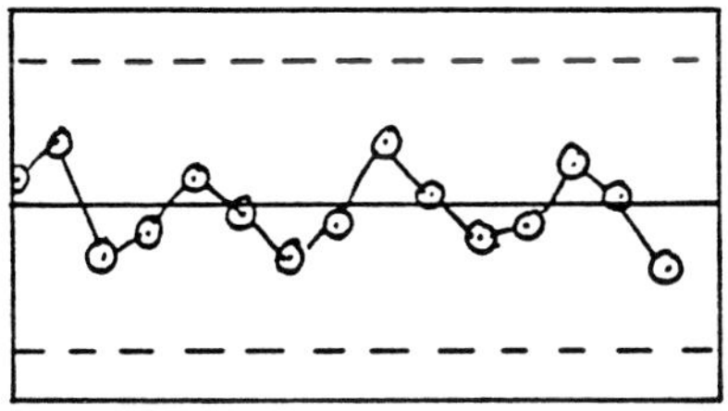

Figure 3-7 Cycle

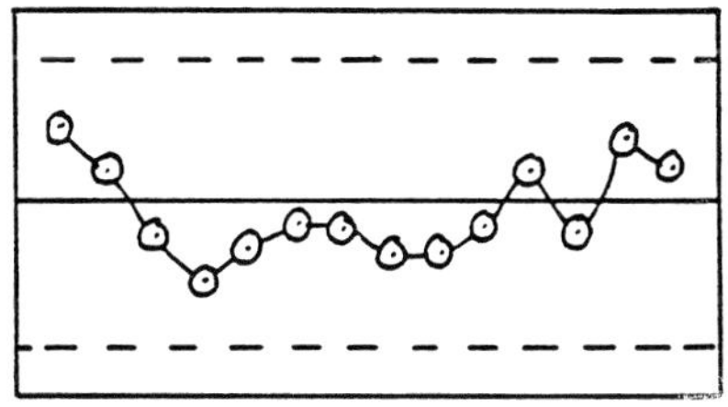

Figure 3-8 Run

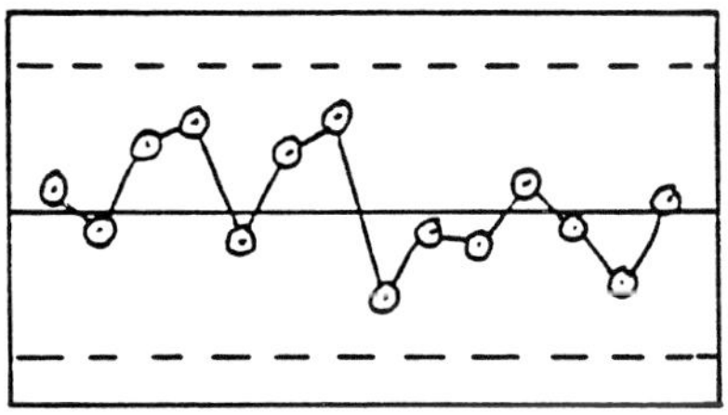

Figure 3-9 Zone rule

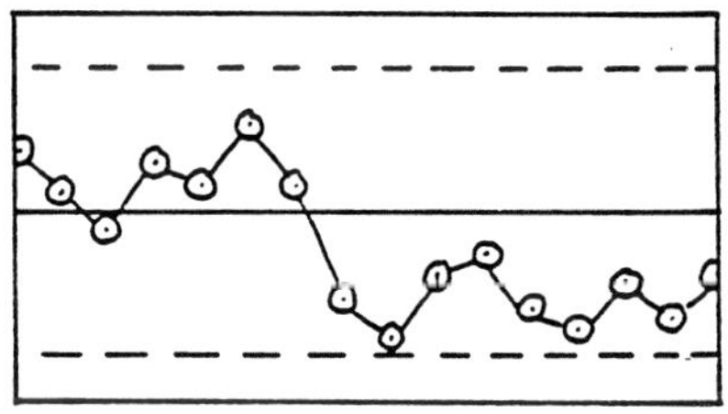

Figure 3-10 Shift in level

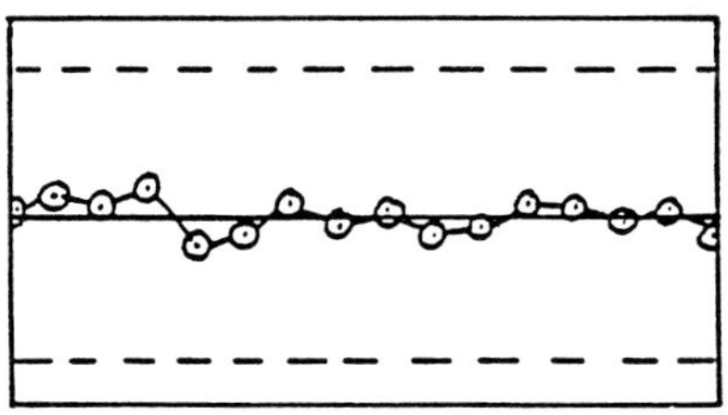

Figure 3-11 Stratification

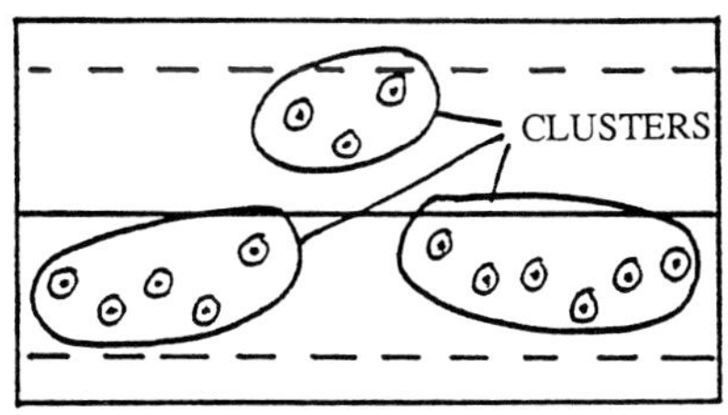

Figure 3-12 Clusters

process is said to be in control. A stable process will have points randomly distributed between the control limits; the $\bar{x}$ & R points should tend not to follow each other. A lack of stability can cause them to move together. A process whose population is positively skewed will cause a positive correlation between the $\bar{x}$ chart and the R chart, see figure 3-14. Conversely, a negatively skewed distribution will cause a negative correlation between the $\bar{x}$ chart and the R chart, see figure 3-15. An example of distributions with positive and negative skew is shown in figure 4-6 of chapter 4.

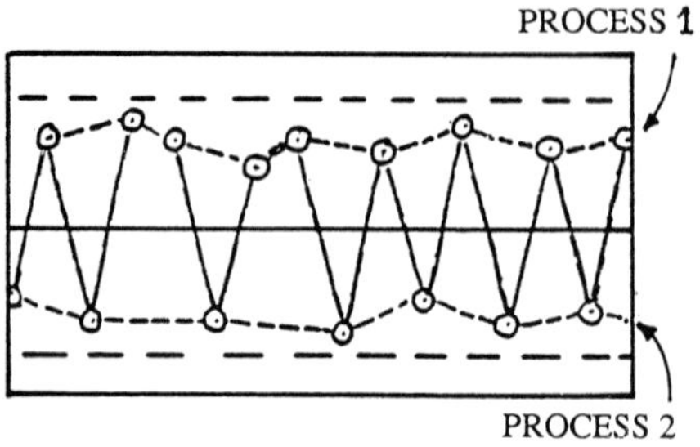

Figure 3-13 Mixture

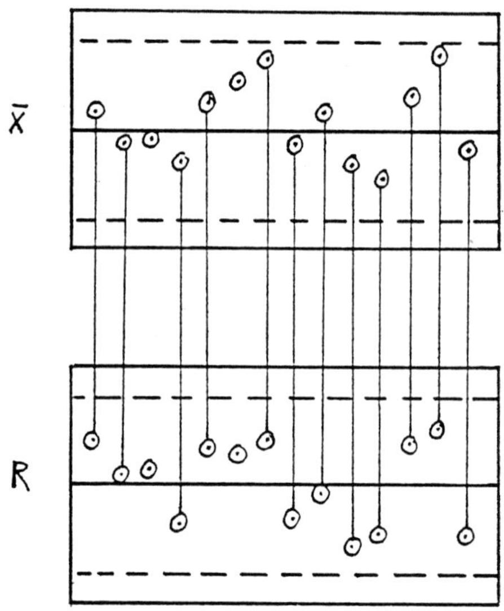

Figure 3-14 $\bar{x}$ & R Chart showing positive correlation

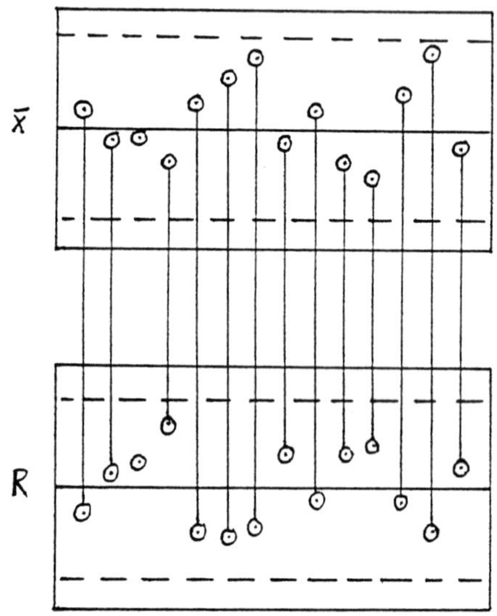

Figure 3-15 $\bar{x}$ & R Chart showing negative correlation

SPC gives us various tools to identify assignable causes of variation. However, the fact that common causes of variation are inherent to the process does not mean that they cannot be reduced or eliminated. Unfortunately, SPC does not help us much in this area. To understand the common causes of variation in a process we have to understand the process technology. Therefore, we will treat extrusion technology in detail so that we can understand the effect of both assignable and common causes of variation and learn how to minimize their effects on the extrusion process.

3.3 CONTROL CHARTS FOR ATTRIBUTES DATA

Attributes data is a count of the number of nonconformities on a unit (number of defects per part) or the number of nonconforming units (defective parts). The following definitions are often used with attributes data:

NONCONFORMITY (DEFECT): Any aspect or fault which causes an item (production unit or product) not to conform to the specified criteria.

c = Number or occurrences of nonconformities in a sample.
u = c/n = Number of occurrences of nonconformities per production unit.
n = Number of production units in the sample.

NONCONFORMING ITEM (DEFECTIVE): An item which does not meet measurement specifications or which contains one or more nonconformities (defects)

np = Number of nonconforming items in a sample of n items.

p = np/n = Proportion of nonconforming items in a sample.

There are four types of control charts for attributes. Control charts for the count of nonconforming units are p-charts and np-charts. For p-charts the sample size varies, for np-charts the sample size is constant. Control charts for the count of individual nonconformities or defects on a product are c-charts and u-charts. For u-charts the sample size varies, for c-charts the sample size is constant. This is shown in Table 3-3.

TABLE 3-3. CONTROL CHARTS FOR ATTRIBUTES DATA

Sample or Subgroup Size	What is Counted	Type of Control Chart
variable (>50)	defective items	p-chart
constant (>50)	defective items	np-chart
constant	defects on an item	c-chart
variable	defects on an item	u-chart

It should be noted that the calculation of the three sigma control limits for attributes is different from that of variables. For variables it is based on the normal (Gaussian) distribution. For p-charts and np-charts it is based on the binomial distribution. For c-charts and u-charts it is based on the Poisson distribution.

3.3.1 CREATING AND ANALYZING ATTRIBUTES CONTROL CHARTS

As with the $\bar{x}$ & R chart, the creation and analysis of attributes charts requires several steps. These are:

Step 1. Select the subgroup period and size. The most effective subgroup size for the p and np-charts is greater than 50. For c and u-charts the subgroup size should be at least one, better between 5 and 10. Subgroups should be chosen to ensure minimum variation within a subgroup.
Step 2. Record the data on the control chart. Within a subgroup, items being inspected should be collected randomly so that each item has an equal chance of having nonconformities.
Step 3. Plot the data.
Step 4. Calculate and draw the control lines. At least 25 subgroups should be used to determine the control limits.
Step 5. Analyze the chart for signs of assignable causes of variation.
Step 6. Eliminate assignable causes and create a new chart with new control limits.
Step 7. If the process is considered in control, the control limits can be extended forward in time. As a process continues to improve the old control limits may become too wide. In this case it can be desirable to determine new control limits based on the most recent data. The narrower the control limits can be made, the higher the quality.

3.3.2 THE p-CHART

The p-chart (percent defective chart) is an attributes chart for the percentage of defective items in a subgroup when the subgroup size is not necessarily constant. The following symbols are used on a p-chart:
n = number of items in a subgroup.
k = number of subgroups in the study period.
np = number of defective items in a subgroup.
p = fraction defective in a subgroup. These are the points that are plotted on the chart. The fraction p is determined from:

$$p = \frac{np}{n} \qquad (3\text{-}1)$$

$\bar{p}$ = The average fraction defective for the study period. This is the central line on the chart. It is determined from:

$$\bar{p} = \frac{np_1 + np_2 + np_3 + \ldots\ldots\ldots\ldots + np_k}{n_1 + n_2 + n_3 + \ldots\ldots\ldots\ldots + n_k} \qquad (3\text{-}2)$$

Since the value of n varies from subgroup to subgroup, the central limits have to be calculated for each subgroup. This makes the chart difficult to make and to read. This can be avoided by keeping the subgroup size the same or within plus or minus 25% of the average sample size and using an average n. It is also possible to show two sets of control limits based on the minimum and maximum subgroup sizes.

The upper control limit is determined using the following expression:

$$UCL_{\bar{p}} = \bar{p} + 3\sqrt{\frac{\bar{p}(1-\bar{p})}{\bar{n}}} \tag{3-3}$$

The lower control limit is determined from:

$$LCL_{\bar{p}} = \bar{p} - 3\sqrt{\frac{\bar{p}(1-\bar{p})}{\bar{n}}} \tag{3-4}$$

where $\bar{n}$ is the average number of items in a subgroup.

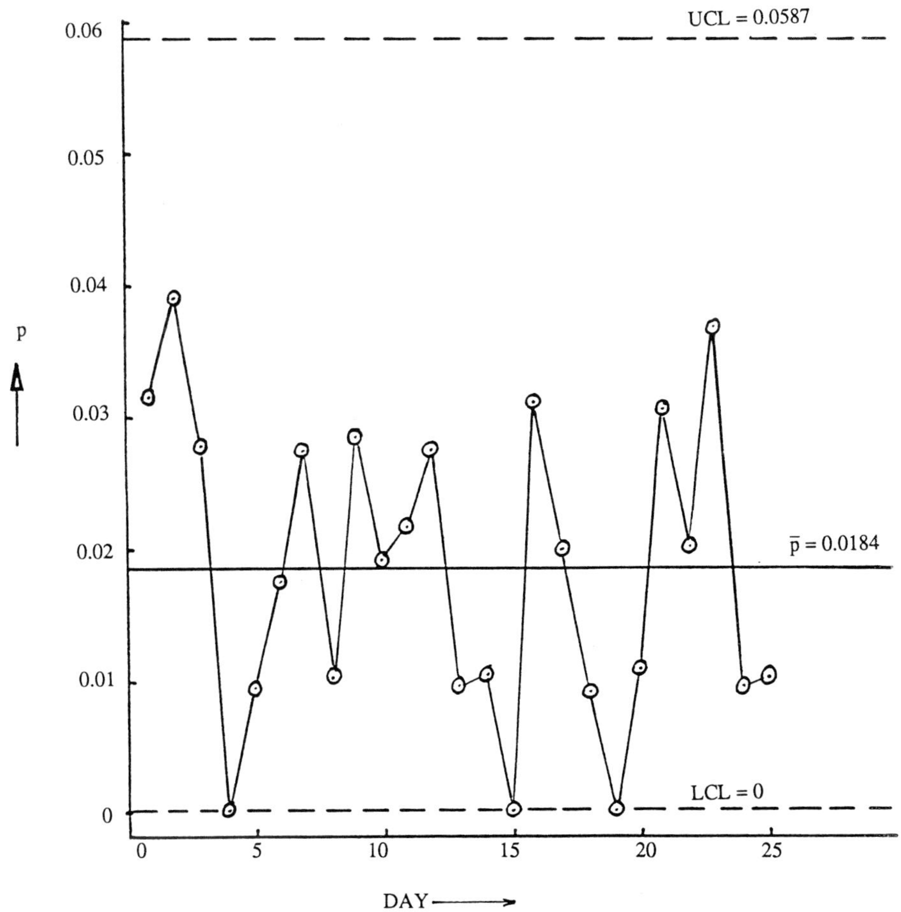

Figure 3-16 p Chart for fraction of defective extruded pieces

An example of a p-chart is shown in figure 3-16. Here the percent defective extruded pieces of pipe is plotted for a 25 day period. The tabular data are shown in table 3-4. For this example the average number of items in a subgroup is:

$$\bar{n} = \frac{2498}{25} = 100$$

The average fraction defective is:

$$\bar{p} = \frac{46}{2498} = 0.0184$$

The upper control limit is:

$$UCL = 0.0184 + 3\sqrt{\frac{0.0184(1-0.0184)}{100}} = 0.0587$$

The lower control limit is:

$$LCL = 0.0184 - 3\sqrt{\frac{0.0184(1-0.0184)}{100}} = -0.0219 \qquad \text{(take zero).}$$

TABLE 3-4. DATA ON DEFECTIVE EXTRUDED PIPES

Day	Subgroup Size, n	Defectives, np	Fraction Defective
1	95	3	0.0316
2	102	4	0.0392
3	108	3	0.0278
4	90	0	0.0000
5	104	1	0.0096
6	112	2	0.0179
7	88	2	0.0273
8	96	1	0.0104
9	105	3	0.0286
10	104	2	0.0192
11	92	2	0.0217
12	109	3	0.0275
13	101	1	0.0099
14	97	1	0.0103
15	93	0	0.0000
16	96	3	0.0312
17	100	2	0.0200
18	107	1	0.0093
19	103	0	0.0000
20	91	1	0.0110
21	97	3	0.0309
22	99	2	0.0202
23	108	4	0.0370
24	103	1	0.0097
25	98	1	0.0102
Sum	$\Sigma n = 2498$	$\Sigma np = 46$	

The methods for finding signs of assignable causes are basically the same as those for $\bar{x}$ & R charts discussed earlier. The p-chart in figure 3-16 shows no clear signs of assignable causes, which indicates that the process is in control.

3.3.3 THE np-CHART

The np-chart is a chart for the actual number of defective items in the subgroup. It requires a constant subgroup size. The average number of defective or non conforming units is the central line on the chart. It is determined from:

$$n\bar{p} = \frac{np_1 + np_2 + np_3 + \ldots\ldots\ldots\ldots + np_k}{k} \quad (3\text{-}5)$$

where k is the number of subgroups.
The upper control limit is determined from:

$$UCL_{np} = n\bar{p} + 3\sqrt{n\bar{p}(1-\bar{p})} \quad (3\text{-}6)$$

where: $\bar{p} = n\bar{p} / n$

The lower control limit is obtained from:

$$LCL_{np} = n\bar{p} - 3\sqrt{n\bar{p}(1-\bar{p})} \quad (3\text{-}7)$$

An example of a np-chart is shown in figure 3-17. Here the number of defective pieces of extruded catheter tubing is plotted over a 24 hour period. The tabular data are shown in table 3-5. For this example the average number of defective units is:

$$n\bar{p} = 96 / 24 = 4.00$$

The average fraction of defective units is:

$$\bar{p} = 4.00 / 60 = 0.067$$

The upper control limit is:

$$UCL = 4.00 + 3\sqrt{4.00(1-0.067)} = 9.797$$

The lower control limit is:

$$LCL = 4.00 - 3\sqrt{4.00(1-0.067)} = -1.797 \qquad \text{(take zero)}$$

Figure 3-17 shows no clear signs of assignable causes.

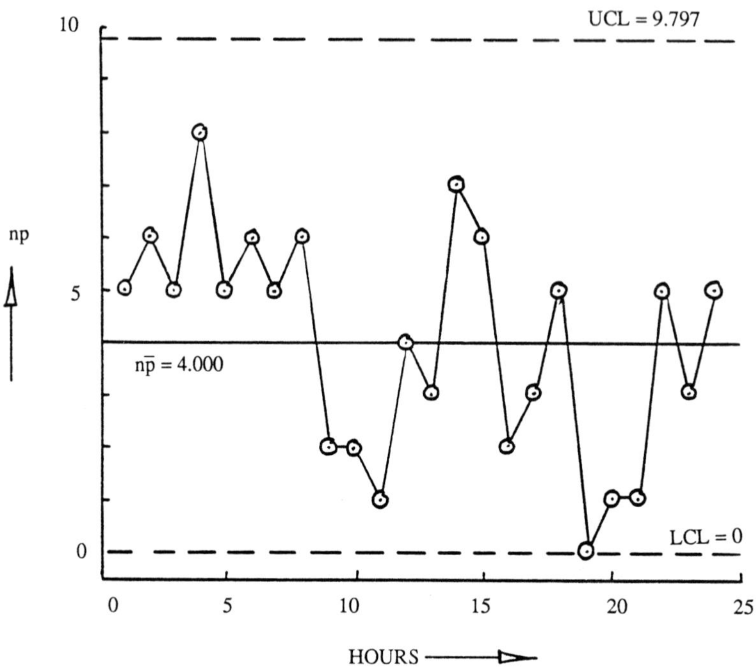

Figure 3-17 np Chart for number of defective pieces of extruded tubing

TABLE 3-5. DATA ON DEFECTIVE PIECES OF EXTRUDED CATHETER TUBING

Hour	Subgroup Size, n	Number Defectives, np
1	60	5
2	60	6
3	60	5
4	60	8
5	60	5
6	60	6
7	60	5
8	60	6
9	60	2
10	60	2
11	60	1
12	60	4
13	60	3
14	60	7
15	60	6
16	60	2
17	60	3
18	60	5
19	60	6
20	60	1

Hour	Subgroup Size, n	Number Defectives, np
21	60	1
22	60	5
23	60	3
24	60	5
		Σnp=96

3.3.4 THE c-CHART

The c-chart shows the number of defects or nonconformities. It is useful in situations where a unit can contain many defects. The c-chart requires a constant subgroup size. It is applied where the defects are scattered continuously throughout the unit output, such as defects in a roll of film, a reel of coated wire, a box of pieces of tubing, etc. Subgroups can be a square foot of film, 10 feet of coated wire, 5 pieces of tubing, etc.

The average number of defects for the study period is determined from:

$$\bar{c} = \frac{c_1 + c_2 + c_3 + \dots\dots\dots\dots\dots + c_k}{k} \tag{3-8}$$

where k is the number of subgroups.

The upper control limit is obtained from:

$$UCL_c = \bar{c} + 3\sqrt{\bar{c}} \tag{3-9}$$

The lower control limit is determined from:

$$LCL_c = \bar{c} - 3\sqrt{\bar{c}} \tag{3-10}$$

Figure 3-18 shows a c-chart for the number of flaws in extruded polyethylene film. A one square foot sample of the film is examined every 1000 feet and the total number of flaws per square foot sample is recorded and graphed. The tabular data is shown in table 3-6. For this example the average number of defects is:

$$\bar{c} = 125 / 25 = 5.00$$

The upper control limit is:

$$UCL = 5.00 + 3\sqrt{5.00} = 11.71$$

The lower control limit is:

$$LCL = 5.00 - 3\sqrt{5.00} = -1.71 \quad \text{(take zero)}$$

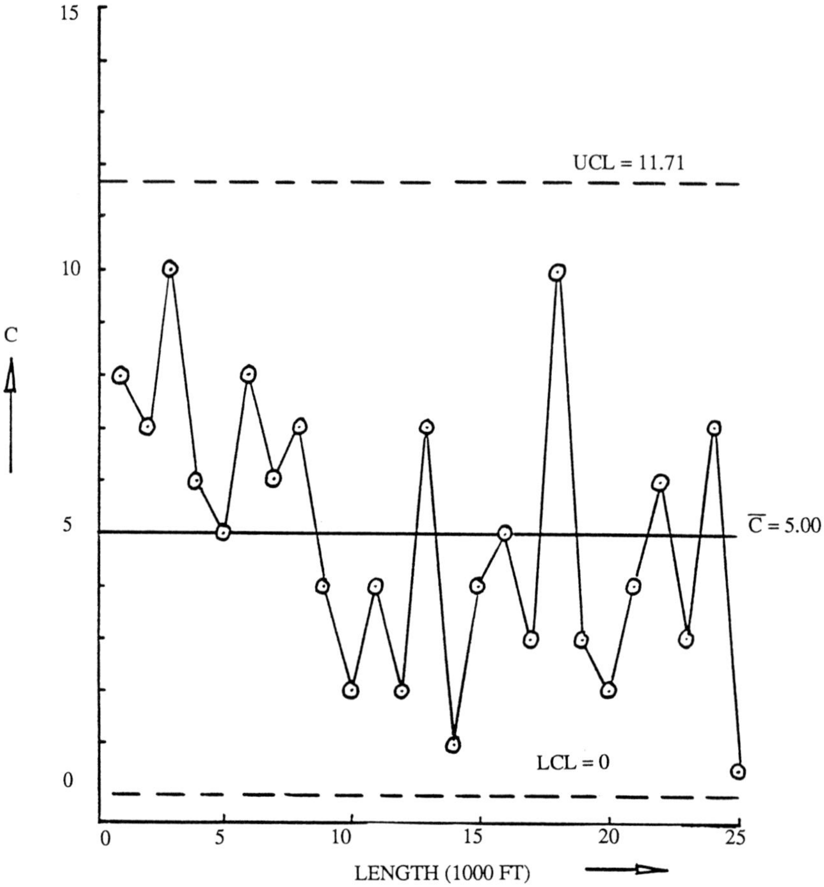

Figure 3-18 c-Chart for number of flaws/ft^2 in pe film

TABLE 3-6. FLAWS IN EXTRUDED FILM

Length	Number of Flaws, c	Length	Number of Flaws, c
1,000	8		
2,000	7	14,000	1
3,000	10	15,000	4
4,000	6	16,000	5
5,000	5	17,000	3
6,000	8	18,000	10
7,000	6	19,000	3
8,000	7	20,000	2
9,000	4	21,000	4
10,000	2	22,000	6
11,000	4	23,000	3
12,000	2	24,000	7
13,000	7	25,000	1
			$\Sigma c=125$

3.3.5 THE u-CHART

The u-chart can be used in the same situation as the c-chart. It must be used when the subgroup size varies. The average number of defects per subgroup is:

$$u = c / n \tag{3-11}$$

where n is the number of units in a subgroup.

The grand average number of defects per unit is:

$$\bar{u} = \frac{c_1 + c_2 + c_3 + \ldots\ldots\ldots\ldots + c_k}{n_1 + n_2 + n_3 + \ldots\ldots\ldots\ldots + n_k} \tag{3-12}$$

where k is the number of subgroups.
The upper control limit is determined from:

$$UCL_u = \bar{u} + 3\sqrt{\bar{u} / \bar{n}} \tag{3-13}$$

where n̄ is the average number of units in a subgroup.

The lower control limit is:

$$LCL_u = \bar{u} - 3\sqrt{\bar{u} / \bar{n}} \tag{3-14}$$

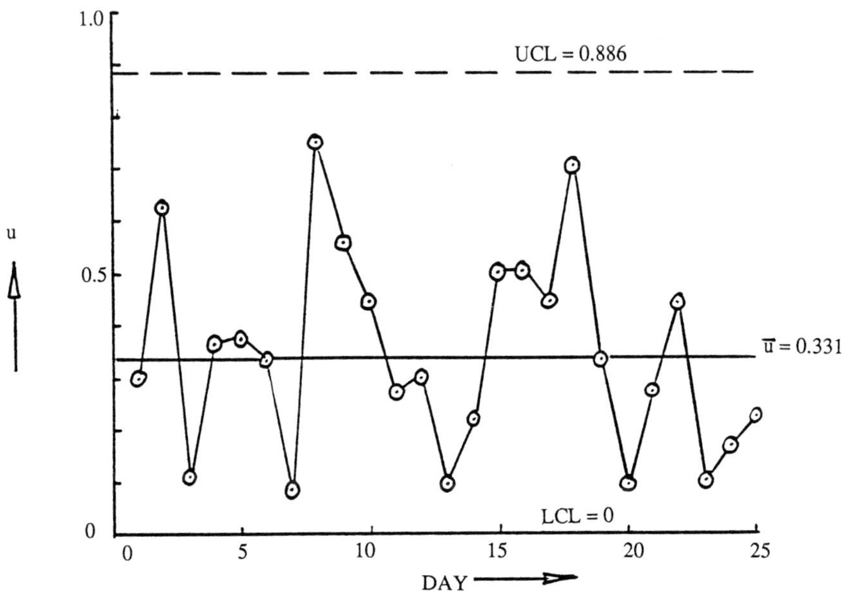

Figure 3-19 u-Chart for number of lumps in tubing

Figure 3-19 shows a u-chart for the number of lumps in pieces of extruded tubing. The number of pieces, n, inspected varies from day to day. As long as n does not vary more than 25% from the average n, the average sample size can be used. The tabular data is shown in table 3-7.

TABLE 3-7.

Day	Sample Size, n	Number of Lumps, c	Defects in Subgroup, u
1	10	3	0.300
2	8	5	0.625
3	9	1	0.111
4	11	4	0.364
5	8	3	0.375
6	9	3	0.333
7	12	1	0.083
8	8	6	0.750
9	9	5	0.556
10	9	4	0.444
11	11	3	0.273
12	10	3	0.300
13	11	1	0.091
14	9	2	0.222
15	10	5	0.500
16	8	4	0.500
17	9	4	0.444
18	10	7	0.700
19	9	3	0.333
20	11	1	0.091
21	11	3	0.273
22	9	4	0.444
23	10	1	0.100
24	12	2	0.167
25	9	2	0.222
	242	80	

For this example, the grand average number of defects per unit is

$$\bar{u} = 80 / 242 = 0.331$$

The average sample size is:

$$\bar{n} = 242 / 25 = 9.68$$

The upper control limit is:

$$UCL = 0.331 + 3\sqrt{0.331 / 9.68} = 0.886$$

The lower control limit is:

$$LCL = 0.331 - 3\sqrt{0.331 / 9.68} = -0.224 \quad \text{(take zero)}$$

4 PROCESS CAPABILITY

4.1 INTRODUCTION

Process capability is a measure of how capable a process is of making parts that are within specifications. Figure 4-1 shows a capable process where the spread of the process data is narrower than the tolerance or specification width. In other words, all the process data are within the upper and lower specification limits. Before the process capability is determined, one has to make sure that the process is in control. This is done by making sure that no assignable causes are affecting the process. Thus, if assignable causes are acting on the process, these have to be eliminated first. In determining process capability we need to determine the spread of the individual process data (x), not of the averages ($\bar{x}$).Therefore, we cannot simply use the control limits from a $\bar{x}$ chart, because the control limits describe the $6\sigma_{\bar{x}}$ spread of $\bar{x}$. As discussed earlier, the standard deviation of the individual data, σ_x , is related to the standard deviation of the averages $\sigma_{\bar{x}}$ by:

$$\sigma_x = \sqrt{n}\sigma_{\bar{x}} \qquad (4\text{-}1)$$

where n is the number of data from which $\bar{x}$ is calculated (the size of the subgroup). The spread of the individual data is always larger than the spread of the averages, see figure 4-2.

Process capability is determined to find out whether a process can meet specifications and how many parts can be expected to be out of tolerance. Capability studies can also be useful in the following situations:

- Evaluation of new equipment purchases
- Equipment selection for production of certain parts
- Setting specifications
- Costing out contracts

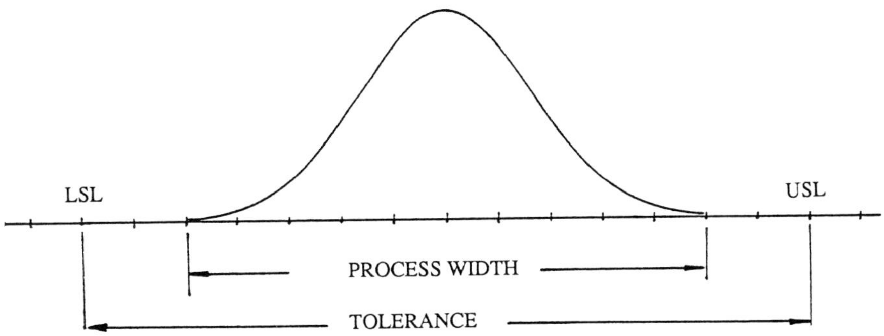

Figure 4-1 Example of a capable process

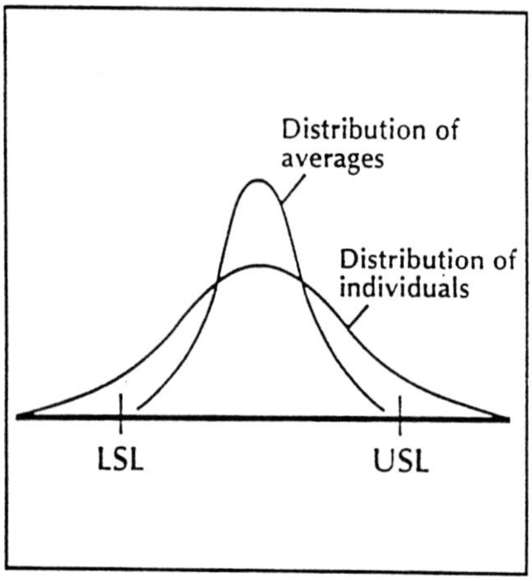

Figure 4-2 Distribution of averages versus the distribution of individuals

The steps involved in a capability study are the following, see figure 4-3:

1. Collect process data
2. Plot the data on control charts
3. Determine the control limits
4. Make sure the process is in control
5. Determine the process capability
6. If the process is not capable, improve the process and return to step 1.

Before process capability is determined, one has to make sure that the process is **in control** and that the individual process data is **normally distributed**. It should be noted that a process can be in control and yet have its individual process data not normally distributed. This happens when the process is inherently non-normal. Certain measurements will always result in non-normal distributions. Examples are roundness, flatness, etc; these measurements have a natural barrier at zero. A perfect measurement is zero; values less than zero are not possible. Standard capability indices are not valid for such non-normal distributions.

4.2 CAPABILITY INDICES

The most commonly used capability indices are the process potential index, CP, and the capability index CpK. The process potential index is the ratio of the tolerance to 6 times the process standard deviation:

$$CP = \frac{TOLERANCE}{6\sigma_x} \tag{4-2}$$

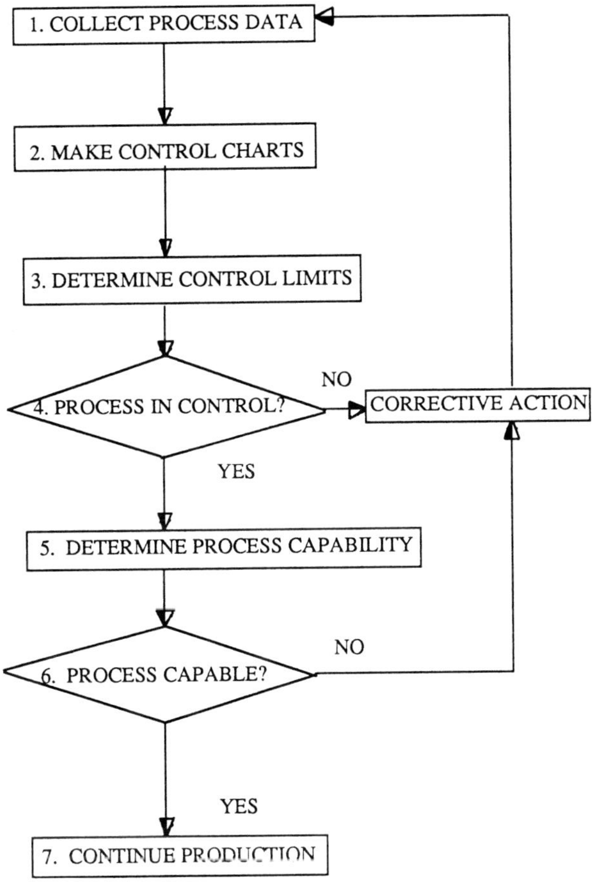

Figure 4-3 Flow chart of process capability study

where: TOLERANCE = USL-LSL. The acronym USL stands for the upper specification limit and LSL is the lower specification limit.

In a normal distribution 99.73% of the data will be within a $6\sigma_x$ spread. Thus, when CP=1.0, it can be expected that 99.73% of the parts will be within the tolerance (assuming that the data is centered around the target value). The higher the value of CP, the more parts will fall within the specification limits. For example, when CP=1.33, it can be expected that 99.994% of the parts will be within tolerance. In other words, only 6 out of 100,000 parts would be expected to be out of tolerance. When CP=1.0, we would expect 270 out of 100,000 parts to be out of tolerance. Thus, a small change in CP can translate in a large change in the number of rejects.

CP is only a measure of spread of the process data. It does not take into account whether or not the data is centered around the target value. If the data is not centered around the target value, using the CP can result in a false sense of security. In this case, the CpK is more useful because it takes both the spread and the center of the data into account.

The expression for CpK is:

$$CpK = \frac{USL - MEAN}{3\sigma_x} \tag{4-3 a}$$

or

$$CpK = \frac{MEAN - LSL}{3\sigma_x} \tag{4-3 b}$$

The CpK is the lowest of these two values. If the mean is larger than the target value, expression 4-3a will yield the lowest CpK. If the mean is smaller than the target value, expression 4-3b will give the lowest CpK.

If the data are centered around the target value, the CP = CpK. However, when the data are not centered around the target value, the CpK will always be smaller than the CP. Figure 4-4 shows examples of CP and CpK values for five different distributions. Many companies require a CpK index of 1.33 to 2.0 for the product from their suppliers. A CpK of 1.33 means that 6 or less parts per 100,000 are expected to be out of specifications. A CpK of 2.0 means that 2 or less parts per 1,000,000,000 are expected to be out of specification. Clearly, a CpK of 2.0 represents an almost negligible chance of rejects.

Sometimes another capability index is used; this is the CR index. The CR index is simply the inverse of CP:

$$CR = \frac{6\sigma_x}{TOLERANCE} \tag{4-4}$$

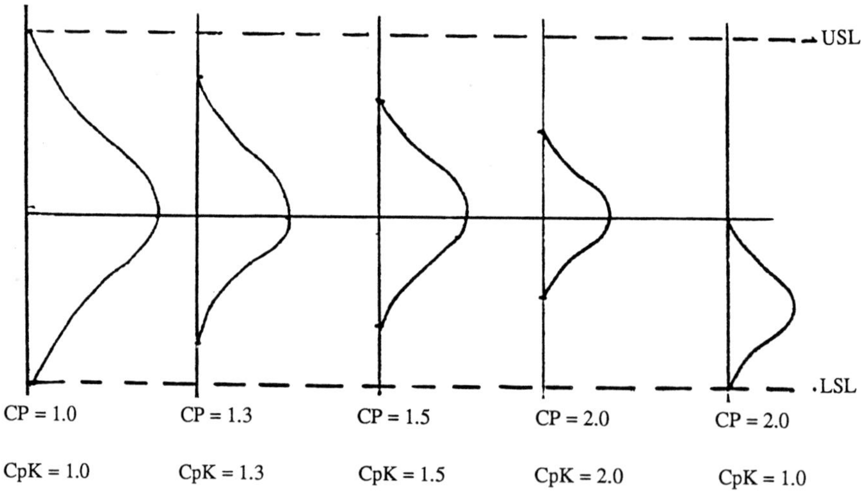

Figure 4-4 CP and CpK values for five distributions

The smaller the CR value the better. The CP index is used more often than the CR index, because it is easier to compare to CpK. Sometimes not only the lowest CpK value is used from expressions 4-3a and 4-3b, but also the largest. The larger of the two CpK values is also referred to as the Zmax/3 index. Using the same nomenclature, the CpK index can also be called the Zmin/3 index.

4.3 TESTS FOR NORMALITY

The capability indices CP, CpK, and CR all assume normal distribution of the data. A number of tests can be used to determine whether or not the distribution is normal. These tests can be divided into two categories, graphical methods and statistical calculations.

GRAPHICAL METHODS

If the sample is large enough, a histogram can be plotted. Then a normal curve with the same mean and standard deviation can be plotted with the histogram to see how

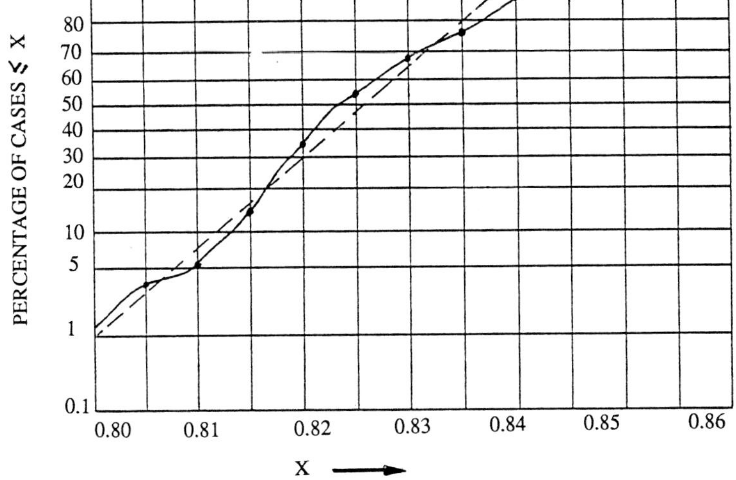

Figure 4-5 Example of probability plot

well the normal curve fits the actual distribution. This is not a quantitative comparison and, therefore, not a good test of normality.

Another graphical method is a plot of the cumulated distribution on normal probability paper to determine how well it is fitted by a straight line. A straight line indicates a normal distribution. This plot is usually referred to as a normal probability plot. Figure 4-5 shows an example of a normal probability plot. There is a reasonable fit to a straight line, however, whether the deviation from a straight line is significant cannot be determined from the graph.

STATISTICAL CALCULATIONS

A commonly used test of normality is the chi-square, χ^2, test. This test is also called goodness-of-fit test, because we determine how well the normal curve fits our data. The data are divided into classes (intervals). Each class has an upper and lower boundary and we determine how many data points fall in each class. If there are not at least five data points in a class, it is combined with the next class. Then the chi-square calculation is performed for each class. The chi-square formula is:

$$\chi^2 = \Sigma \frac{(O - E)^2}{E} \tag{4-5}$$

Where O is the observed frequency and E the expected frequency based on the normal distribution. The Σ means that the $(O-E)^2/E$ values for all classes are summed to obtain chi-square. This value is compared to the value of the chi-square table 4-1.

DF stands for the number of degrees of freedom. It is determined by the number of classes minus three. We subtract three because we are estimating three statistics; mean, standard deviation, and frequency. The different columns represent different factors. Typically, a risk factor of 0.95 is used. The value from the table is called the critical chi-square. If the calculated chi-square value is smaller than the critical value based on a risk factor of 0.95, we can say that the distribution is normal with a better than 95% confidence level. Further information on the chi-square test can be found in reference 3.

Other calculations to test for normality determine certain measures of non-normal characteristics such as skewness and kurtosis. If a distribution is symmetrical (zero skewness), its third moment about the mean will be zero. The third moment or skewness is a measure of how much heavier the distribution is on one side versus the other. It is determined from:

$$M_3 = \frac{\Sigma(x_i - \bar{x})^3}{n - 1} \tag{4-6}$$

TABLE 4-1. THE χ^2 DISTRIBUTION TABLE

DF	χ^2.005	χ^2.01	χ^2.05	χ^2.10	χ^2.90	χ^2.95	χ^2.99	χ^2.995
1	39E-6	16E-5	39E-4	0.016	2.71	3.84	6.63	7.88
2	0.010	0.020	0.103	0.210	4.61	5.99	9.21	10.60
3	0.072	0.115	0.352	0.584	6.25	7.81	11.34	12.84
4	0.207	0.297	0.711	1.064	7.78	9.49	13.28	14.86
5	0.412	0.554	1.15	1.61	9.24	11.07	15.09	16.75
6	0.676	0.872	1.64	2.20	10.64	12.59	16.81	18.55
7	0.989	1.24	2.17	2.83	12.02	14.07	18.48	20.28
8	1.34	1.65	2.73	3.49	13.36	15.51	20.09	21.96
9	1.73	2.09	3.33	4.17	14.68	16.92	21.67	23.59
10	2.16	2.56	3.94	4.87	15.99	18.31	23.21	25.19
11	2.60	3.05	4.57	5.58	17.28	19.68	24.73	26.76
12	3.07	3.57	5.23	6.30	18.55	21.03	26.22	28.30
13	3.57	4.11	5.89	7.04	19.81	22.36	27.69	29.82
14	4.07	4.66	6.57	7.79	21.06	23.68	29.14	31.32
15	4.60	5.23	7.26	8.55	22.31	25.00	30.58	32.80
16	5.14	5.81	7.96	9.31	23.54	26.30	32.00	34.27
18	6.26	7.01	9.39	10.86	25.99	28.87	34.81	37.16
20	7.43	8.26	10.85	12.44	28.41	31.41	37.57	40.00
24	9.89	10.86	13.85	15.66	33.20	36.42	42.98	45.56
30	13.79	14.95	18.49	20.60	40.26	43.77	50.89	53.67
40	20.71	22.16	26.51	29.05	51.81	55.76	63.69	66.77
60	35.53	37.48	43.19	46.46	74.40	79.08	88.38	91.95
120	83.85	86.92	95.70	100.6	140.2	146.6	158.9	163.6

If the distribution has a long tail to the right, the third moment about the mean will have a positive value. A negatively skewed distribution will have a negative value of the third moment about the mean. Figure 4-6 shows a positively and negatively skewed distribution. Skewness may be estimated by the difference between the mean and the mode divided by the standard deviation:

$$M_3 \approx \frac{\bar{x} - \hat{x}}{s} \tag{4-7}$$

Further information can be found in Duncan's book (reference 3).

A measure of kurtosis is the ratio of the fourth moment to the second moment square about the mean; the fourth moment about the mean is determined from:

$$M_4 = \frac{\Sigma(x_i - \bar{x})^4}{n - 1} \tag{4-8}$$

For a normal distribution this M_4/M_2^2 ratio has a value of 3. When the ratio is larger than 3, the distribution has a sharper peak and fatter tails than the normal distribution. When the ratio is less than 3, the distribution has a flatter peak and thinner tails than the normal distribution.

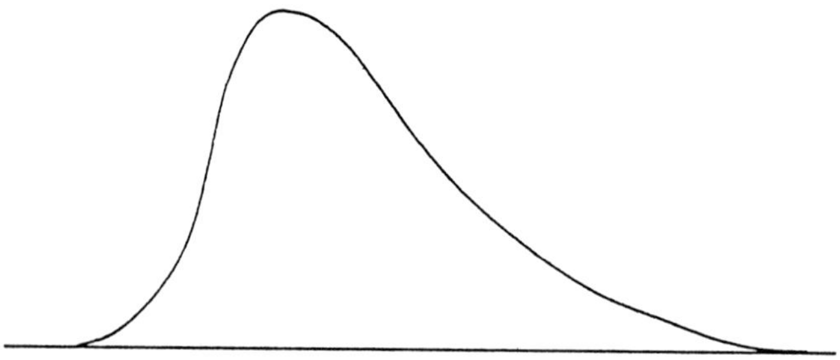

DISTRIBUTION WITH POSITIVE SKEW

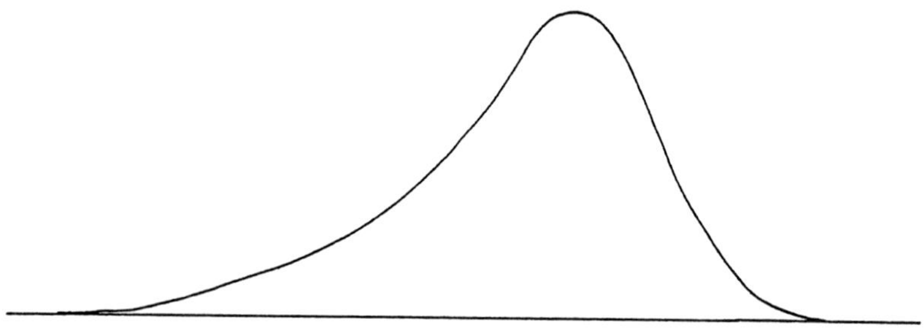

DISTRIBUTION WITH NEGATIVE SKEW

Figure 4-6 Examples of positive and negative skew

4.4 CAPABILITY MEASURES FOR NON-NORMAL DISTRIBUTIONS

If the distribution that we are dealing with is not normal, there are various approaches to determine the capability of the process. In some cases the non-normal distribution can be transformed to a normal distribution. This is usually done by taking the natural logarithm of the data, including the specification limits. If the distribution has a natural barrier at zero, the process capability can be expressed as the percentage of parts between zero and the upper specification limit. Another option is to calculate the CpK for just the upper half of the distribution.

If the distribution is unimodal but not symmetrical, see figure 4-7, we can split the distribution data in two halves. The central point should be determined by the mode. Each half is plotted separately and its mirror image added to give two approximately normal distributions. Each distribution curve can than be analyzed for the percentage of parts outside the appropriate specification limit. In figure 4-7, we should calculate the area below LSL using curve I and the area above USL using curve II. Since we are only concerned with half of each distribution, the area within the curve is determined by subtracting the area outside the specification limit from

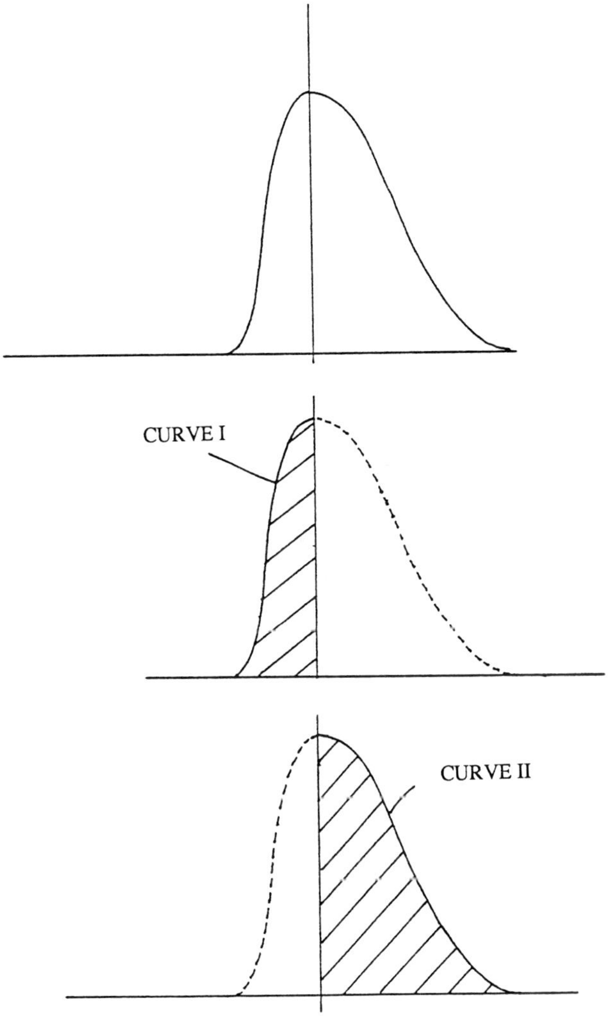

Figure 4-7 Splitting of a unimodal but asymmetrical distribution

0.50. The total percentage of the actual distribution within specification limits is found by adding the results of half of curve I and half of curve II. This total percentage can be used to quantify the process capability. A condition for this approach, obviously, is that both halves of the curves can be fitted reasonably well with a normal curve.

Several other methods to determine process capability for non-normal distributions are available. One is based one the Weibull distribution. Another useful method is the use of statistical tolerances according to Wilks (12). This approach is valid for any distribution as long as the variable is continuous and the sampling random. Wilks derived an expression from which one can calculate the sample size n necessary for an assurance A that at least 100B percent of the contents of a lot or the output of a random process will be included between the largest and smallest values in a random sample from the lot or process:

$$nB^{n-1} - (n-1)B^{n} = 1 - A \qquad (4\text{-}9)$$

The problem with the Wilks method is that for a high level of assurance, e.g. A = 0.95, and a high value of B, e.g. 100B = 99.73%, the sample size becomes quite large. For A = 0.95 and B = 0.9973, the necessary sample size, n, is 1755. This method, therefore, may not be practical if the measurements are made by hand. However, if measurements are made automatically and on-line, e.g. a diameter measurement using a laser gage, then a large sample size is no major problem. Even if the data are collected only once per second, then a sample of 1755 measurements will be collected within half an hour.

Clements (24) proposed a simple method for calculating CP and CpK for any shape distribution, using the Pearson family of curves. The calculation requires estimates of the mean, standard deviation, and kurtosis for the process. The advantages of this method are:

- When the distribution is normal the indexes are exactly the same as those given by the traditional method.
- The only difference from the traditional procedure is the method of calculating the width and position of the top and bottom halves of the distribution relative to the tolerance limits.
- It does not require mathematical transformation of the data.
- It is easy to visualize graphically and explain to non-statisticians.
- It is relatively easy to calculate manually or by using a hand-held calculator.

5 MEASUREMENT

5.1 INTRODUCTION

When a process is analyzed the observed variation is often assumed to be the same as the actual variation. There is, however, an important component in the observed variation which is the variation due to the measurement. When the measurement variation is large, the observed variation can be substantially different from the true variation. In the extreme situation, the measurement variation can be so large that it overshadows the actual variation. Obviously, in this case accurate process analysis is not possible. Therefore, accurate process analysis requires a determination of the variation due to measurement to make sure that this is only a small part of the observed variation.

Measuring is the process of quantification, comparing an unknown magnitude to a known magnitude. The science of measurement is called ***metrology.*** To quantify data on parts or processes, defined standard units must be used. These are called ***units of measure.*** Several systems of international units have been developed. The most important one is the SI system, which is an acronym for the French words: Système International d'Unités. The SI system uses meter as the basic measurement of length, Newton for force, second for time, and kilogram for mass. Most industrialized countries have adopted the SI system with the exception of the United States. In the U.S. the English system is still widely used. This uses the basic measurement of foot for length, pound for force, and second for time. However, increasing international contacts are bringing about an increased use of the SI system in the U.S.

Any practical system of measurement standards should be based on units that are unchangeable. Primary reference standards are maintained at the National Bureau of Standards. In the U.S. this bureau is called the National Institute of Standards and Technology. The standards consist of copies of the international kilogram and measuring systems which can verify the units and subunits of the defined standards. The primary reference standards are used for calibration of measuring instruments. Since the primary reference standards cannot be used for calibration of all instruments, secondary and tertiary standards were developed. These allow transfer of the primary standards for calibration of instruments in laboratories and manufacturing areas. Measuring instruments used by operators, technicians, process engineers, etc. are calibrated against a set of working standards. The working standards are referred back to the primary standards through transfer standards. A precision of about five to one is required to transfer from one standard to the next. When an instrument is calibrated such that it can be related to a primary standard, the calibration is said to have traceability. The hierarchy of standards is shown in figure 5-1.

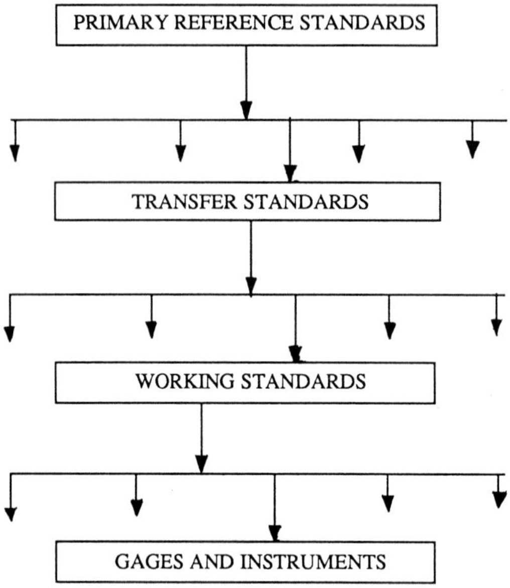

Figure 5-1 The hierarchy of standards

5.2 BASIC CONCEPTS

Various terms are used to describe the characteristics of measuring instruments. Two important terms are accuracy and precision. These terms are often used interchangeably. However, they have distinctly different meaning in describing measuring instruments.

Accuracy is the extent to which the average of many repetitive measurements on a single item agree with the true value. The difference between good accuracy and poor accuracy is illustrated in figure 5-2. An accurate instrument may or may not be precise.

Precision is the extent to which repetitive measurements on a single unit agree. The difference between good precision and poor precision is illustrated in figure 5-3. It should be noted that the true value is not shown in figure 5-3, because the true value is not needed to evaluate precision.

The term bias is related to accuracy. ***Bias*** is the difference between the true value and the average of repetitive measurements made by a measurement process. When the bias is large the measuring instrument is said to be out of calibration. The ***sensitivity*** of a measuring instrument is the smallest change in the measured quantity which the instrument is capable of detecting. ***Resolution*** of a measuring instrument is the smallest unit of measure which an instrument is capable of indicating. Sensitivity and resolution are related terms, however, they have different meanings.

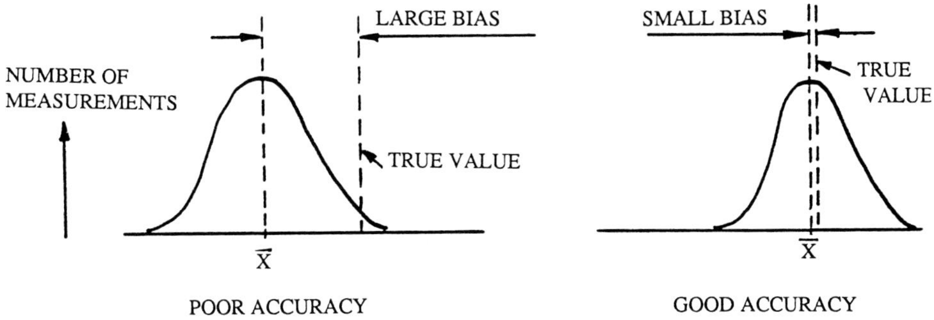

Figure 5-2 Illustrations of poor and good quality

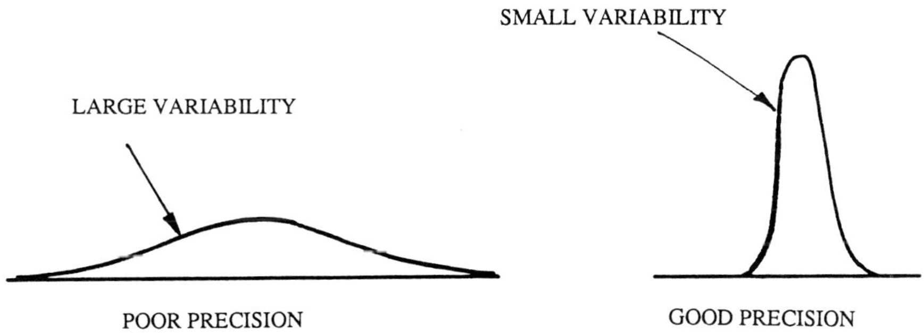

Figure 5-3 Illustrations of poor and good precision

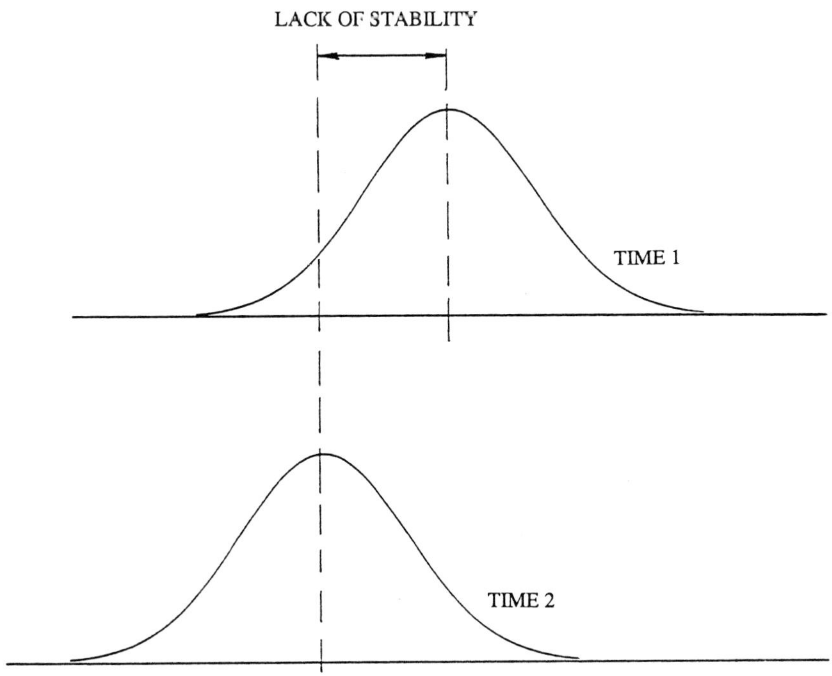

Figure 5-4 Illustration of lack of stability

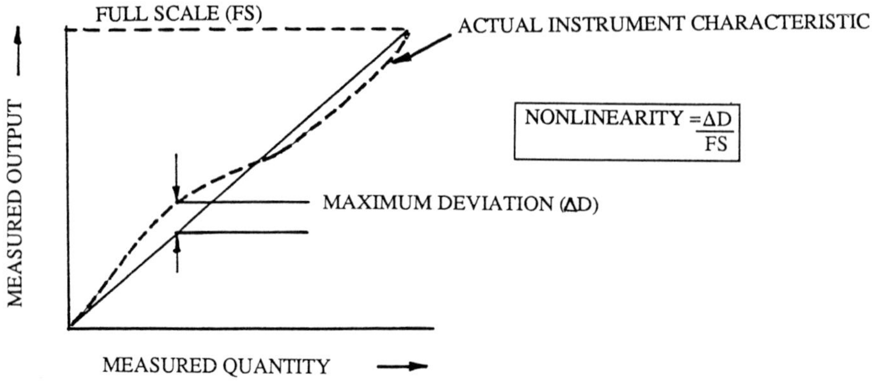

Figure 5-5 Illustration of non-linearity

Sensitivity describes the ability of a measuring instrument to detect or sense small changes, while resolution describes the ability to indicate or display small changes. ***Stability*** refers to the difference in the average of at least two sets of measurements obtained with the same measuring device on a single unit taken at different times. This is illustrated in figure 5-4. Stability is often expressed as a percentage of full scale. An ideal instrument has an absolutely linear relationship between the measured quantity and the measured output. The deviation from linear or straight line behavior is described by the term non-linearity. ***Non-linearity*** is the largest deviation of the actual measurements from the straight line characteristic divided by the full scale, see figure 5-5.

5.3 MEASUREMENT ERROR

Measurement error is the difference between the measured value and the true value. It can be due to a problem of accuracy and/or precision. Various sources of error play a role in most instruments: non-linearity, deformation, hysteresis, poor electrical connections, and sensitivity to environmental factors such as temperature, humidity, magnetic fields, electrical fields, etc. Another important factor can be operator error. When two operators measure the same product with the same measuring instrument there will be differences in the recorded values. Differences can also occur when different test procedures are used to measure the same product. Transposition errors can occur when data has to be read and recorded by the operator. If the data are not recorded legibly transcription errors can occur later when the data is analyzed.

To keep measurement error very small it is recommended that the ratio of product tolerance to the precision of the measurement instrument be at least 5:1 and prefera-

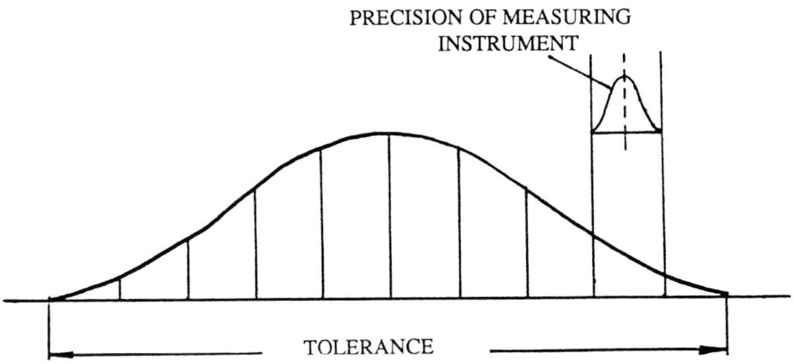

Figure 5-6 Measurement precision should be 10% of product tolerance

bly 10:1. This is illustrated in figure 5-6. The observed variation is a combination of true product or process variation and measurement variation. This relationship can be expressed as:

$$\sigma_{obs} = \sqrt{\sigma^2_{meas} + \sigma^2_{true}} \quad (5\text{-}1)$$

In this expression σ_{obs} is the standard deviation of the observed data, σ_{meas} the standard deviation due to measurement, and σ_{true} the actual standard deviation of the product.

To minimize measuring error, instruments should be calibrated regularly. Calibration control should provide for periodic audits on instruments to check their accuracy, precision, and general condition. The calibration schedule for an instrument should be kept at the instrument to show how long ago the instrument was calibrated. It is recommended to keep control charts on auditing tools. Tool control charts allows current process data to be accurately compared to historical data or to correct data for possible bias. Tool control charts also provide an indication of the need for calibration or repair.

Operator error can be reduced by training operators to use uniform and consistent measuring procedures. These procedures should be documented and provide enough detail to prevent different measuring techniques. Operator error can often be largely eliminated by automating the measurement. Instruments that automatically record their measurements will eliminate transposition and transcription errors. An additional advantage of automated measurements is speed, which means that more measurements can be made in the same period of time. Therefore, it is recommended to use automated measurements as much as possible.

5.4 QUANTIFYING MEASUREMENT VARIATION

All measurements are subject to variability. For accurate process analysis it is necessary to determine whether the variability due to measurement is only a small fraction of the observed variability. Detailed procedures have been developed to assess the measurement error quantitatively. For a more complete discussion on the subject the reader is referred to the Measurement System Analysis Manual, reference 17. The degree of precision in a measurement process is generally designated as the standard deviation of measurement error. Common cause variation contributing to the precision is a combination of variation due to repeatability and reproducibility.

Repeatability is the variation obtained when one person measures the same quantity several times using the same measuring instrument. See figure 5-7.

Reproducibility is the variation in measurement averages due to differences among instruments, operators, etc., see figure 5-8.

The standard deviation of the repeatability can be estimated from the average range. In control chart applications the relationship is:

$$\hat{\sigma}_{rpt} = \overline{R} / d_2 \qquad (5\text{-}2)$$

Constant d_2 is given in table 5-1.

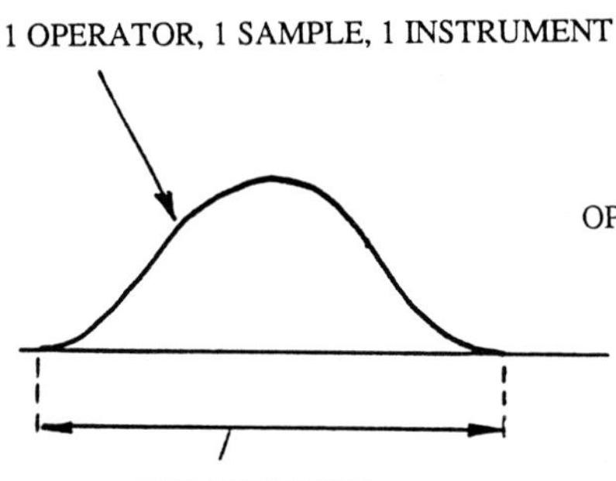

Figure 5-7 Illustration of repeatability

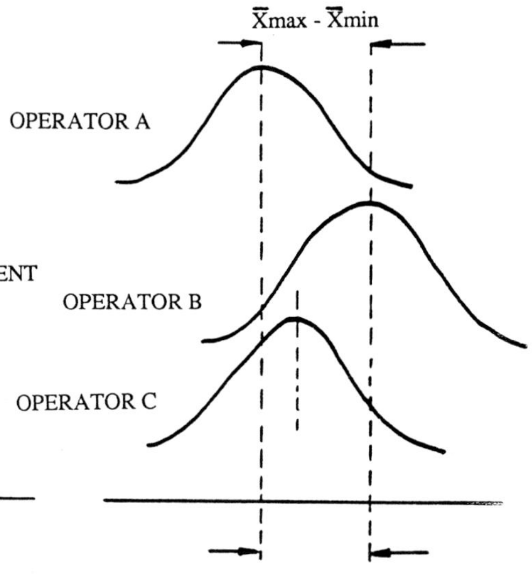

Figure 5-8 Illustration of reproducibility

TABLE 5-1.

	n=1	2	3	4	5	8	10	>10
k=2	0.709	0.781	0.813	0.826	0.840	0.855	0.862	0.885
3	0.524	0.552	0.565	0.571	0.575	0.581	0.581	0.592
4	0.446	0.465	0.472	0.474	0.476	0.481	0.481	0.485
5	0.403	0.417	0.420	0.422	0.424	0.426	0.427	0.429
6	0.375	0.385	0.388	0.389	0.391	0.392	0.392	0.395
7	0.353	0.361	0.364	0.365	0.366	0.368	0.368	0.370
8	0.338	0.344	0.346	0.347	0.348	0.348	0.350	0.351
9	0.325	0.331	0.332	0.333	0.334	0.334	0.336	0.337
10	0.314	0.319	0.322	0.323	0.323	0.324	0.324	0.325

n is the number of parts measured (number of samples).
k is the number of times each part was measured (sample size).

Table 5-1. Factors $1/d_2$ for converting the average Range, $\bar{R}$, into estimated standard deviation $\hat{\sigma}_x$. Courtesy reference 3.

The estimated standard deviation of the reproducibility is determined from:

$$\hat{\sigma}_{rpd} = (\bar{x}_{max} - \bar{x}_{min}) / d_2 \qquad (5\text{-}3)$$

Constant d_2 is determined by the number of parts measured (n) and the number of times each part was measured (k). The standard deviation of the measurement error is obtained from the relationship:

$$\sigma_c = \sqrt{\sigma_{rpt}^2 + \sigma_{rpd}^2} \qquad (5\text{-}4)$$

The total measurement variation is a combination of variation due to precision, accuracy, and stability. This is illustrated in figure 5-9. In general, a measurement pro-

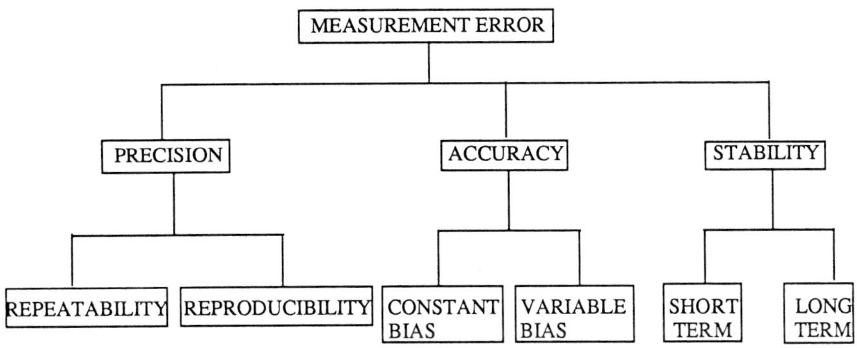

MAJOR COMPONENTS OF MEASUREMENT ERROR

Figure 5-9 Components of measurement error

cess is considered capable when it is stable and its measurement error does not account for a significant portion of the tolerance band. The short term precision should be less than 10% of the tolerance:

$$\frac{6\sigma_e}{USL - LSL} < 0.10 \tag{5-5}$$

where σ_e is the short term precision.
Short term is typically expressed in hours, while long term is expressed in days, weeks, or even months. The long term precision should be less than 25% of the tolerance:

$$\frac{6\sigma_{el}}{USL - LSL} < 0.25 \tag{5-6}$$

where σ_{el} is the long term precision.

A number of different procedures have been developed over the years to analyze measurement systems. A collection of Institute of Standards papers (4) describes procedures developed at this organization. Another useful reference is the Western Electric Statistical Quality Control Handbook (5). The procedure for analyzing measurement systems described in this handbook is often referred to as the "Western Electric" method. Another good reference is the General Motors Statistical Process Control Manual (6). In the following, we will discuss the methods described in the GM manual in more detail.

5.4.1 SHORT METHOD FOR GAGE R&R

Gage R&R refers to repeatability and reproducibility. This method requires a set of five parts to be measured by two operators. Each part is measured only once. A drawback of the short method is that repeatability cannot be isolated from reproducibility. Thus, the results reflect a combination of both types of measurement error.

The parts are selected at random and should have different values. The results have to be recorded on a table as shown in table 5-2. The difference between measurements of operator A and B is entered in the column labeled "Range (A-B). Only positive numbers are entered. The average range, $\bar{R}$, is the sum of all ranges divided by the number of parts. In the example in table 5-2, the average range $\bar{R}$=1.4.

The gage error is calculated by multiplying the average range by a constant 4.33. With $\bar{R}$=1.4, the result is GR&R=6.1, where GR&R stands for gage repeatability and reproducibility and represents the combined gage error. To convert this number to a percentage of tolerance, divide by the tolerance and multiply by 100. If the tolerance for the example above is 20, then the GR&R(%) becomes:

$$\text{GR\&R(\% tolerance)} = \frac{6.1 \times 100}{20} = 30.5\%$$

TABLE 5-2. DATA COLLECTION SHEET

Parts	Operator A	Operator B	Range (A-B)
1	4	2	2
2	3	4	1
3	6	7	1
4	5	7	2
5	9	8	1
		Sum of Ranges	7

average range ($\bar{R}$) = $\Sigma R/5 = 7/5 = 1.4$
gage error (GR&R) = $4.33 x \bar{R} = 4.33 x 1.4 = 6.1$

This level of gage error is generally considered to be too high. If the process variability is known, the GR&R can be compared to the process variation rather than the tolerance. For example, if the difference between the upper and lower control limit is 12, then GR&R as a percentage of the 6σ process variation becomes:

$$GR\&R(\%\ 6\sigma_{process}) = \frac{6.1 \times 100}{12} = 50.8\%$$

Again, this level of gage error is considered too high in most cases.

5.4.2 LONG METHOD FOR GAGE R&R

This method allows separate determination of gage repeatability and reproducibility. The results of the study can also provide information on the causes of gage error.

If lack of reproducibility is large compared to repeatability, possible causes can be:

- The operator may not be properly trained in how to use and read the gage instrument.
- The calibrations on the gage dial may not be clear.

If lack of repeatability is large compared to reproducibility, the reasons may be:

- The gage instrument needs maintenance.
- A better gage may be required.
- The location of the part in the gage may not be optimum.

The number of operators, the number of trials, and the number of parts may be varied. However, the study should be conducted according to the following steps:

1. Refer to operators A, B, C, etc. The parts should be numbered so that the operator cannot identify the number. This increases the chance that all parts are measured the same way. A typical number of parts is 10.

2. Calibrate the gage.
3. Let operator A measure all parts in random order and enter the results in column 1 of data collection sheet, see table 5-3.
4. Repeat step 3 with operator B, C, etc.
5. Repeat steps 3 and 4 for the number of trials required.

TABLE 5-3. GAGE REPEATABILITY AND REPRODUCIBILITY DATA SHEET

	1	2	3	4	5	6	7	8	9	10	11	12
	Operator A				Operator B				Operator C			
Sample	Trial 1	Trial 2	Trial 3	range	Trial 1	Trial 2	Trial 3	range	Trial 1	Trial 2	Trial 3	range
1	0.55	0.60	0.55	0.05	0.50	0.55	0.55	0.05	0.55	.050	0.55	0.05
2	0.95	1.00	0.95	0.05	1.00	0.95	1.05	0.10	1.05	1.00	0.95	0.10
3	0.90	0.90	0.85	0.05	0.75	0.75	0.75	0.0	0.75	0.85	0.80	0.10
4	0.90	0.90	0.90	0.0	0.70	0.75	0.80	0.10	0.75	0.80	0.80	0.05
5	0.60	0.55	0.55	0.05	0.40	0.40	0.45	0.05	0.50	0.55	0.55	0.05
6	0.95	1.05	0.95	0.10	0.95	1.05	1.00	0.10	1.05	1.05	1.00	0.05
7	1.00	1.00	0.95	0.05	0.90	0.90	0.95	0.05	0.95	0.95	1.00	0.05
8	0.90	0.80	0.90	0.10	0.80	0.70	0.75	0.10	0.80	0.85	0.80	0.05
9	0.95	1.00	0.95	0.05	0.90	0.90	0.95	0.05	1.00	1.00	1.05	0.05
10	0.55	0.65	0.60	0.10	0.45	0.50	0.55	0.10	0.90	0.85	0.85	0.05
total	8.25	8.45	8.15	0.60	7.35	7.45	7.80	0.70	8.30	8.40	8.35	0.60
	↳	8.25		$\bar{R}_a$=	↳	7.35		$\bar{R}_b$=	↳	8.30		$\bar{R}_c$=
		8.15	↲	0.06		7.80	↲	0.07		8.35	↲	0.06
	Sum=24.85				Sum=22.60				Sum=25.05			
	$\bar{x}_a$=0.8283				$\bar{x}_b$=0.7533				$\bar{x}_c$=0.8350			

The average range is determined as follows:

$$\bar{R} = (\bar{R}_a + \bar{R}_b + \bar{R}_c) / 3 = (0.06 + 0.07 + 0.06) / 3 = 0.0633$$

With constant D_4=2.574, see table 3-2, the upper control limit for the range is:

$$UCL_R = \bar{R} x D_4 = 0.0633 x 2.574 = 0.163$$

The difference between the maximum $\bar{x}$ and minimum $\bar{x}$ is:

$$\Delta \bar{x} = \bar{x}_{max} - \bar{x}_{min} = 0.8350 - 0.7533 = 0.0817$$

NOTES:

6. Steps 3 through 5 may be modified for large size parts, unavailability of parts, or when operators are on different shifts.
7. Enter the observations on a data collection sheet as shown in table 5-3 and calculate gage R&R using the formulas on the gage R&R report such as shown in table 5-4.

TABLE 5-4. GAGE REPEATABILITY AND REPRODUCIBILITY REPORT

Name and Part Number:	Gasket, GA-104
Characteristic:	PTFE
Specification:	0.6-1.0 mm
Gage Name:	Thickness Gage
Gage number:	X-2034
Gage Type:	0-10 mm
Date (month-day-year):	6-15-92
Performed by:	M. Schneringer

From the data sheet: $\bar{R}=0.0633$ and $\Delta\bar{x}=0.0817$

MEASUREMENT UNIT ANALYSIS

REPEATABILITY - EQUIPMENT VARIATION (E.V.)

$E.V.=\bar{R}xK_1=0.0633x3.05=0.193$

K_1(two trials)=4.56 and K_1(three trials)=3.05

%E.V.=100xE.V./TOLERANCE=100x0.193/0.40=48.25%

REPRODUCIBILITY - APPRAISER VARIATION (A.V.)

$$A.V.=\sqrt{[(\Delta\bar{x})\ x\ K_2]^2-(E.V.)^2/(nxr)}=\sqrt{(0.0817x2.70)^2-0.193^2/(10x3)}=0.2055$$

where n is the number of parts, n=10, and r is the number of trials, r=3;
K_2(two trials)=3.65 and K_2(three trials)=2.70
When the value under the square root sign is negative, the appraiser variation defaults to zero, A.V.=0

%A.V.=100xA.V./TOLERANCE=100x0.2055/0.40=51.38%

REPEATABILITY AND REPRODUCIBILITY (R & R)

$$R\&R=\sqrt{(E.V.)^2+(A.V.)^2}=\sqrt{0.193^2+0.2055^2}=0.282$$

$$\%R\&R=\sqrt{(\%E.V.)^2+(\%A.V.)^2}=\sqrt{48.25^2+51.38^2}=70.48\%$$

Note: All calculations are based on predicting 5.15σ, which corresponds to 99% of the area under the normal curve.

GAGE ACCURACY

Accuracy is generally not as much of a problem as precision. To determine gage accuracy, the true measurement of the sample parts is needed. This can often be done with tool room or layout inspection equipment. A true average is determined from these readings. This is compared later to the operators observed averages ($\bar{x}_A$, $\bar{x}_B$, $\bar{x}_C$, etc.)

If the sample parts cannot be measured this way, the following method can be used. Measure one of the sample parts precisely on tool room or layout inspection equipment. Have one operator measure the same part at least 10 times, using the gage being evaluated. The difference between the true measurement and the observed averages represents gage accuracy. To convert accuracy to a percentage of tolerance, multiply by 100 and divide by the tolerance.

If there is significant lack of accuracy, possible causes can be:

- Error in the master.
- Worn gage.
- Improper calibration.
- Improper use of gage by operator.
- Gage made to the wrong dimension.
- Gage is measuring the wrong characteristic.

GAGE STABILITY

Gage stability should be determined although it usually is not as much of a factor as repeatability and reproducibility. How gage stability is determined depends on how often the gage is used between normal calibrations. The need for calibration depends on factors such as time, the number of measurements taken, air pressure change, warm-up, etc. Given that these factors are known, the calibration frequency can be established to minimize gage error due to stability.

If a gage is used intermittently, then its stability can be determined at the same time as the gage R&R study is made. The gage must be calibrated before and after each trial to determine the amount of calibration change. This amount is the gage stability for that trial only. To determine the overall gage stability, the calibration change for each trial must be added up and divided by the number of trials:

$$\text{GAGE STABILITY} = \frac{\text{SUM OF CALIBRATION CHANGES}}{\text{NUMBER OF TRIALS}}$$

If the gage is normally used for relatively long periods of time without calibration, the stability can be determined without recalibration for each trial. In this case, another gage R&R study is conducted just before the time calibration is due. Gage stability is then the absolute value of the difference between the grand averages of all the measurements in the first and second study.

To convert gage stability to a percentage of tolerance, multiply by 100 and divide by the tolerance. If gage stability error is large, check for these possible causes:

- The gage is not calibrated as often as necessary.
- Air pressure regulator of filter may be needed for air gaging
- Warm-up may be required for electronic gages.

GAGE LINEARITY

Gage linearity is assessed by determining the gage accuracy through the expected operating range. At least two accuracy studies should be conducted, one at each end of the operating range. The gage linearity is:

GAGE LINEARITY = LARGEST ACCURACY VALUE – SMALLEST ACCURACY VALUE

If the gage linearity error is large, check for the following possible causes:

- Gage not calibrated properly.
- Error in the minimum or maximum master.
- Gage is worn.
- Gage may have inherently non-linear characteristics.

Guidelines

The criteria for acceptance of gage repeatability and reproducibility are:

GR&R (% tolerance) < 10% – Acceptable.

GR&R (% tolerance) = 10%-30% – Possibly acceptable, depending on importance of application, cost of the gage, cost of repairs, etc.

GR&R (% tolerance) > 30% – Generally not acceptable.

5.5 MEASUREMENTS IN EXTRUSION

The degree of process control that can be attained in the extrusion process is strongly determined by how well measurements are made. This means not only measuring the appropriate process parameters, but also using the right sensors and transducers and placing these in the correct location.

5.5.1 IMPORTANT PROCESS PARAMETERS

In a typical extrusion process there are between about 40 and 80 process parameters that should be monitored to control the process and ensure consistency. The most important parameters, more or less in order of priority, are:

1. Extrudate dimensions.
2. Diehead pressure (after screen pack).
3. Barrel pressure (before screen pack).
4. Polymer melt temperature.
5. Motor load.
6. Screw speed.
7. Take-up speed.
8. Power consumption of various heating zones.
9. Cooling rate at each cooling zone.
10. Barrel and die temperatures.

The two most important measurements are extruded product dimensions and diehead pressure. Obviously, product dimensions have to be monitored to make sure that the product is within dimensional specifications. The preferred method is to measure on-line on a continuous basis, i.e. real time monitoring. If the measurement data is fed to a computer, then ranges, control limits, standard deviations, etc. can be calculated almost instantaneously. Thus, control charts can be generated and displayed within seconds of the actual measurement. This provides real time SPC; this can be very useful in accurate monitoring and control of the process.

5.5.1.1 EXTRUDATE DIMENSIONS

Various measurement techniques are available to determine extruded product dimensions (see reference 7). These are based on laser, optical light, electromagnetic, ultrasound, inductive, and pneumatic-inductive measurements. The first three are used mostly to determine outside dimensions of extruded products; the last four are used to measure wall thickness. Laser gages are frequently used to measure the outside diameter of tubing, pipe, and coated wire. The resolution varies from 0.001mm to 0.01mm, depending on the range.

Wall thickness is often measured with ultrasound gages. This works well on pipe and large diameter tubing, however, on small (diameter less than 10mm) thin walled tubing this measurement is problematic. The ultrasound measurement is basically a measurement of time; it measures how long a sound wave takes to travel from the outside of the wall to the inside and back. It must be realized that the propagation of sound waves not only depends on the type of material, but also on the material temperature. As a result, considerable measurement error can occur when the stock

temperature of the material changes. This can happen when the line speed changes, when extruder temperatures are changed, when the cooling water temperature changes, etc. The measuring head is typically submerged in water to provide good coupling between the head and the product to be measured. Care must be taken that clean water is used without air bubbles present. Air bubbles can cause considerable measurement error.

The choice of the measurement method will depend on several factors, such as the shape and dimensions of the extruded product, the tolerance, the cost of gage, user friendliness, etc. Measurement of the product dimensions is often the most critical measurement in the extrusion process. As a result the gage should be subjected to a careful R&R study to make sure that the gage is acceptable for the application.

5.5.1.2 PRESSURE

The pressure measurement is important for two main reasons, process control and safety. The diehead pressure, i.e. the pressure after the screen pack and the breaker plate, determines the output. A steady pressure will give a steady output, however, a varying pressure will result in a varying output. Output itself is not easily measured in the extrusion process. However pressure can be measured rather easily. It is most important to determine how diehead pressure changes with time. An instantaneous pressure readout is not very meaningful. The variation of pressure with time can be captured on a simple chart recorder or, better, on a CRT when a computerized data acquisition system (DAS) is available.

The change in output, ΔQ, resulting from a change in pressure, ΔP, can be determined from the following relationship:

$$\Delta Q(\%) = \Delta P(\%) \div n$$

In this expression n is the power law index of the material. For polymer melts n varies between 0 and 1; for most polymers the range is 0.3 to 0.7. (See reference 16, page 218). This means that a 2% pressure fluctuation for a polymer with a power law index n=0.4 will result in an output fluctuation of 5%! In other words, the output fluctuation is considerably larger than the pressure fluctuation. This is true for all polymers. Output fluctuations will tend to cause variations in the extruded product dimensions. Thus, it is very important to keep pressure fluctuation as small as possible. If a circular product is extruded, the percent change in the diameter resulting from output variation will be about half the percent output variation. Thus, an 5% output variation will cause approximately a 2.5% diameter variation.

As an example, let us consider the extrusion of a 10 mm diameter rod. The specifications are 10.3-9.7mm; this is 10.0 mm $\pm$ 3%. The maximum output variation that can be tolerated in this case is about 6%; this is twice the percentage of the specifications. If the material is LDPE with a power law index n=1/3, a 2% pressure fluctuation will cause a 6% output fluctuation. In this case, the maximum pressure fluctuation that we can tolerate is 2%. This of course, assumes that there are no other sources of dimensional variations, which is a questionable assumption. Therefore, we would want to keep the pressure fluctuation to less than 2% in order to keep the extruded product within specifications. Clearly, if the actual pressure fluctuation is 5%, there is little or no chance that we can extrude the product within specifications.

The safety aspect of pressure measurement is critically important. Under certain conditions (e.g. cold startup) an extruder can generate dangerously high pressures, much in excess of 70 MPa (10,000 psi). MPa stands for megaPascal, which is 1,000,000 Pascal. One Pascal is one Newton per square meter. If no pressure relief safety is present, such high pressures can force the die from the extruder at high speed. Any person in the path of this die is likely to be severely injured. Unfortunately, a number of serious accidents have happened as a result of excessive pressure buildup. All extruders should be equipped with a pressure relief device, such as a rupture disk (also called blow plug) or a shear pin. Unfortunately, this is not always the case. It is extremely important, therefore, that a properly working pressure measurement is available on every operating extruder, preferable with a high limit alarm and automatic shut down. The most critical location of the pressure transducer for determining excessive pressures is just before (upstream) the breaker plate. Thus, for process control the pressure measured behind the breaker plate is most important, for safety the pressure just before the breaker plate is most important. It is advisable, therefore, to have at least two pressure transducers on an extruder. The difference in the two pressure readings is a good measure of the buildup of contamination on the screen pack.

Most pressure transducers used in extrusion are electronic transducers that use strain gages on a diaphragm to measure pressure. The strain gage diaphragm is placed a considerable distance away from the diaphragm that is actually in contact with the polymer melt. This is done to minimize the effect of the high temperature of the polymer melt. Two types of electronic transducers are used: the capillary type and the pushrod type, see figure 5-10. In the capillary type a fluid is used to hydraulically couple the melt diaphragm and the strain gage diaphragm. The fluid is usually mercury.

A drawback of these electronic transducers is that they are susceptible to damage. In mercury filled transducers this will release mercury in the polymer melt and the work place which is highly undesirable. The damage is due to the thin diaphragm in contact with the melt. A pressure transducer that is not removed before an extruder barrel is cleaned, is likely to be damaged. Since these transducers are not inexpensive, typically around $1000, frequent damage can be costly. In fact, in some instances the pressure measurement is eliminated all together because of this problem. This, however, is dangerous because of the excessive pressures that can occur in extruders. On the other hand, there are piezoresistive pressure transducers that are more rugged and less susceptible to damage. These newer types of pressure tran-

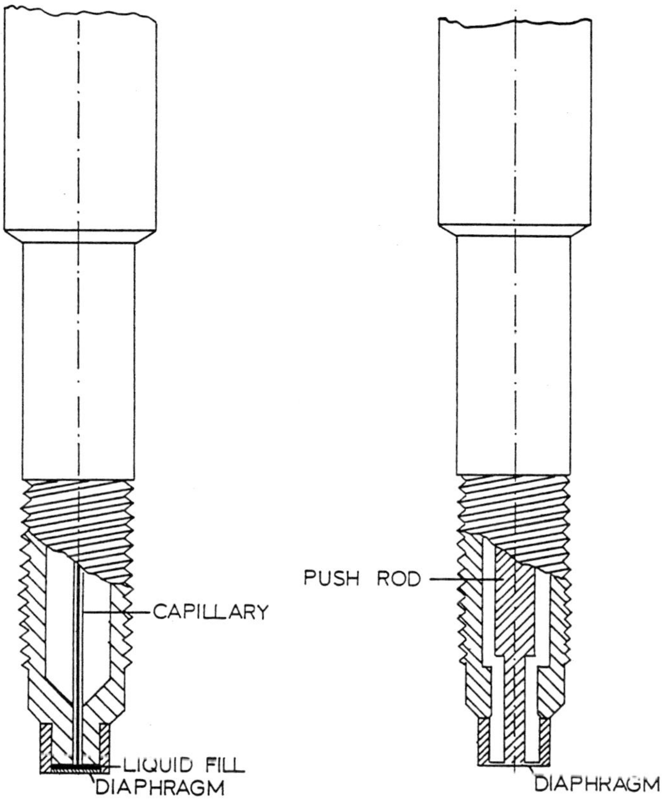

Figure 5-10 Two types of pressure transducers

sducers (see reference 8) feature silicon-on-sapphire technology to obtain good accuracy, stability, and sensitivity over a wide temperature range up to 370°C (700°F).

The latest development in pressure transducers is the optical pressure transducer. This transducer is based on optical intensity modulation (23) and was developed by the German company FOS Meßtechnik in Osterrönfeld. Light from an infrared LED (light emitting diode) is coupled to a fiber optic light guide. The light is guided to the sensor head and the reflected light intensity is compared to the incoming light intensity. As the diaphragm of the transducer is deformed by pressure, the ratio of reflected and incoming light intensity changes and, thus, the pressure is measured. The advantage of this type of transducers is rugged design, good dynamic response, high temperature capability, and small measurement error. Reportedly, the US company Dynisco has also developed an optical pressure transducer and will introduce it the second half of 1992. A comparison of various pressure transducers is shown in table 5-5.

Electronic pressure transducers are sometimes equipped with a temperature sensor. These transducers are typically referred to as melt pressure/temperature (P/T)

TABLE 5-5. COMPARISON OF VARIOUS PRESSURE TRANSDUCERS

Pressure measurement	Reliability/ robustness	Temp. range °C	Temp. sensitivity	Dynamic response	Total error%
Bourdon	Poor	200	Poor	Poor	3–5
Pneumatic	Good	400	Poor	Poor	1.5
Capillary* strain gage	Fair	400	Fair	Fair	0.5–3
Pushrod strain gage	Fair	400	Poor	Fair	3
Piezoelectric**	Good	120	Fair	Excellent	0.5–1.5
Piezoresistive	Good	400	Good	Good	0.2–0.5
Optical	Good	600	Good	Good	0.2–0.5

* Environmental concerns with mercury filled transducers
** Limited to dynamic pressure measurement

transducers. However, it should be realized that the temperature sensor is usually placed a considerable distance from the diaphragm. Therefore, the temperature sensed is not a good indication of the actual melt temperature. There can be considerable differences between the sensed temperature and the actual melt temperature, as much as 50 degrees C (90 degrees F) or more (see reference 9).

For accurate extrusion operations the pressure fluctuation should be kept less than 1%. Assuming that one normally operates at about 1/4 of full scale (FS) pressure, this represents about 0.25% FS. This means that for accurate extrusion operations the pressure measurement error should be less than 0.10% FS. Typical total measurement errors for electronic pressure transducers range from 0.5 to 3.0% FS, which is too much for accurate extrusion operations. Piezoresistive and optical transducers can do better with a total error from 0.2 to 0.5% FS, even though the error is larger than the 0.1% required for accurate extrusion. As a result, piezoresistive and optical transducers will be better suited than electronic transducers. The sensitivity and resolution should be at most 0.1 MPa (14.5 psi), preferably 0.01 MPa (1.45 psi) or less. With a typical full scale of 700 MPa (10,000 psi) this corresponds to 0.145% FS, preferably 0.0145% FS. Commercial gages sometimes have a resolution as high as 2% FS. These are rather useless for accurate process monitoring.

5.5.1.3 TEMPERATURE

The melt temperature is the most important temperature in the extrusion process. It is usually measured at the end of the extruder with an immersion temperature sensor. For a good melt temperature measurement it is important that there is good contact between the polymer melt and the sensor. That is why pressure/temperature

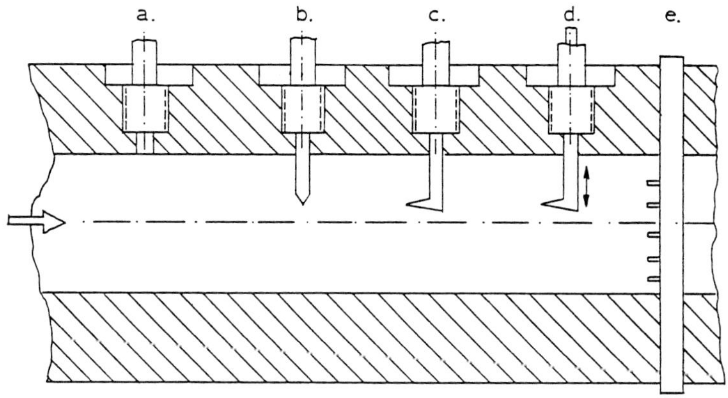

Figure 5-11 Various melt probes; a. Flush mounted, b. Straight protruding, c. Upstream fixed, d. Upstream radially adjustable, e. Bridge with multiple probes

transducers are not suitable for melt temperature measurement, as discussed earlier. Even flush mounted melt temperature sensors are not too useful, because the measured temperature is mostly representative of the metal wall temperature. Various types of melt temperature probes are shown in figure 5-11.

Barrel and die temperatures are also quite important in the extrusion process. These are usually measured with a bayonet-type thermocouple as shown in figure 5-12. Tests (see reference 10) with conventional thermocouples have shown that significant measurement errors can result from:

- Air currents around the extruder can cause errors of up to 50 degrees C (90 degrees F).
- Insufficient depth of the thermocouple (TC) well can cause errors of up to 10 degrees C (18 degrees F). The depth should be at least 25 mm (1 inch).

Special thermocouples have been designed to minimize conduction error. These TCs typically have an exposed sensing element on a ceramic substrate for thermal insulation. An example of such an insulated TC is shown in figure 5-13.

When measuring barrel temperatures, it is important to measure close to the inside diameter of the barrel. After all, it is the polymer temperature that is really important in the extrusion process. It should be realized, however, that in some commercial extruders the temperature sensor is placed close to the outside diameter of the barrel. In fact, in some instances the temperature sensors are placed in the barrel heaters. This will yield good temperature control, however, the **wrong** temperature is being controlled. We should be concerned about the polymer temperature and therefore, we should measure and control the temperature as close to the polymer as reasonably possible.

In analyzing the extrusion process, one should know what type of sensors and transducers are used and where they are located. If this information is not known, entirely

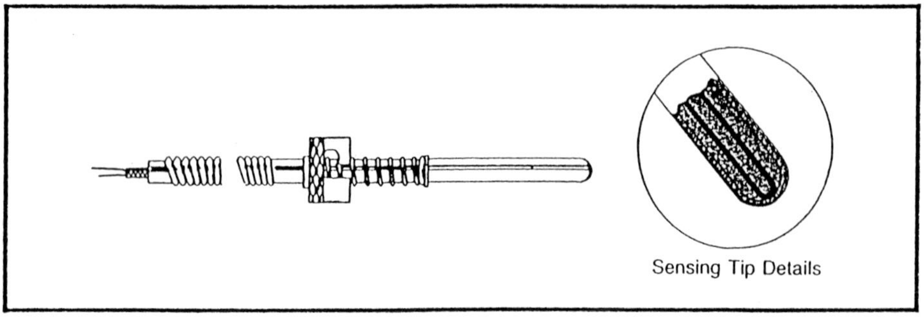

Figure 5-12 Conventional bayonet-type thermocouple

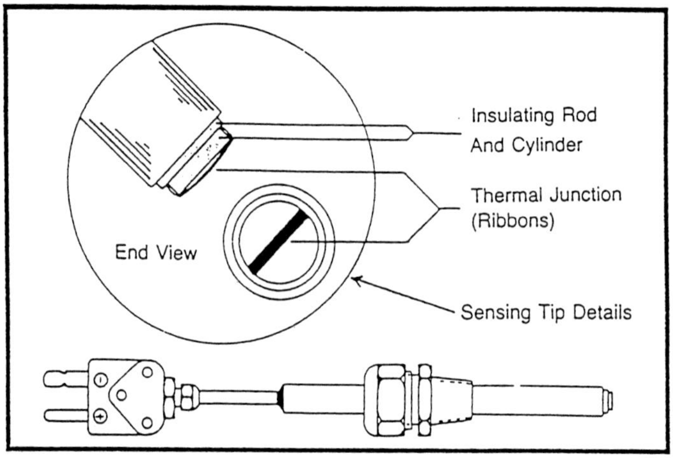

Figure 5-13 Special melt temperature thermocouple

incorrect conclusions could be drawn. For instance, with TCs in the barrel heater, the temperature control would appear to be excellent. At the same time, however, the temperature at the inside barrel could fluctuate wildly, causing major upsets in the extrusion process. The best situation is to have both a deep and shallow barrel temperature measurement, as long as the deep well temperature is the one being controlled. This allows determination of the radial temperature gradient and, thus, the amount of heat being added or removed from the process. There are also dual sensor temperature control systems that use both the shallow and deep temperature sensor. The deep temperature is the one being controlled, however, the shallow temperature participates in the control process through a cascade loop (see reference 11).

5.5.1.4 SCREW AND TAKE-UP SPEED

The output of the extruder is determined directly by the screw speed. It is important, therefore, that the screw speed can be accurately measured and controlled. The screw speed should be measured and displayed to at least 0.1 rpm, preferably 0.01 rpm or even less. In other words, the sensitivity and resolution of the screw speed measurement should be 0.1 rpm or better. Assuming a full scale of 200 rpm, this represents 0.05% FS. This is particularly important when an extruder is operated at low screw speed. For example, if an extruder runs at 5 rpm and the resolution of the screw speed is 1 rpm, the actual screw speed could vary as much as 20% without any indication from the screw speed readout. With a resolution of 0.1 rpm, the actual screw speed could still vary as much as 2% without an indication from the screw speed readout.

Another important consideration is the speed regulation of the extruder drive and the take-up drive. Typical brush type DC drives with tachometer feedback have a speed regulation of 1% FS. Assuming a full speed of 100 rpm, this means that at 10 rpm the screw speed can vary ± 1 rpm. This means ± 10%, a total variation of 20%! This is unacceptable in most extrusion operations. Much better speed regulation can be obtained with digital drives, such as the brushless DC drive or the DC brush with pulse encoder. These drives have a speed regulation of 0.01% FS or better. That means that even at low screw speed, these drives can maintain a very steady screw speed.

Clearly, good speed regulation is equally important for the take-up. If the extruder output is steady but the take-up speed is varying, the extrudate dimensions will vary as well because the draw down of the extruded product will change as the take-up speed changes. It is recommended, therefore, that both the extruder and the take-up device be equipped with drives that provide a speed regulation of 0.01% FS or better. Since digital drives nowadays are no more expensive that the old fashioned DC brush drives with or without tachometer feedback, there is little reason not to use digital drives.

6 EXTRUSION TECHNOLOGY

6.1 INTRODUCTION

Extrusion is a process for the continuous production of plastic products such as tubing, pipe, film, sheet, coated wire, fiber, profiles, etc. Most products have a constant cross section, however, it is also possible to make products with a non-constant cross section. An example of the latter is extrusion blow molding, where a parison is extruded and, subsequently, inflated with air in a mold to make a blow molded product. There are various aspects of the machine design and the method of process operation that strongly affect the stability of the process and the quality of extruded product. The resulting variability, in most cases, is ever present and is, in a statistical sense, predictable. These causes of variation, therefore, are common or inherent causes. SPC focuses primarily on special or unnatural causes. However, it is very important not to stop when special causes of variation have been eliminated. Unfortunately, SPC offers little guidance in the elimination of common cause variations. The most important ingredient in eliminating common cause variation is a good knowledge and understanding of the process technology. SPC is a very useful tool in process analysis, improvement, and control. However, SPC by itself is not enough, it needs to be combined with process technology know-how to really pay dividends.

The extruder commonly used is a single screw extruder. A typical single screw extruder is shown in figure 6-1. The main components of an extruder are:

1. The extruder screw
2. The extruder barrel
3. The feed hopper
4. The feed throat casting
5. The die
6. The drive for the screw
7. Heating and cooling elements
8. Instrumentation and control system

The extruder screw is a long cylinder with one or more helical flights wrapped around it. The screw is located in the extruder barrel. The barrel is a long cylinder with a small clearance between the screw and the barrel. A typical radial clearance is one-thousandth of the barrel diameter. The screw normally rotates in a stationary barrel. The rotation of the screw causes a forward conveying of the plastic from the feed opening to the die exit. When the extruder is fed with solid plastic particles, it is called a **plasticating extruder.** When it is fed with molten plastic, it is called a melt fed extruder. We will deal mostly with plasticating extruders since these are most common.

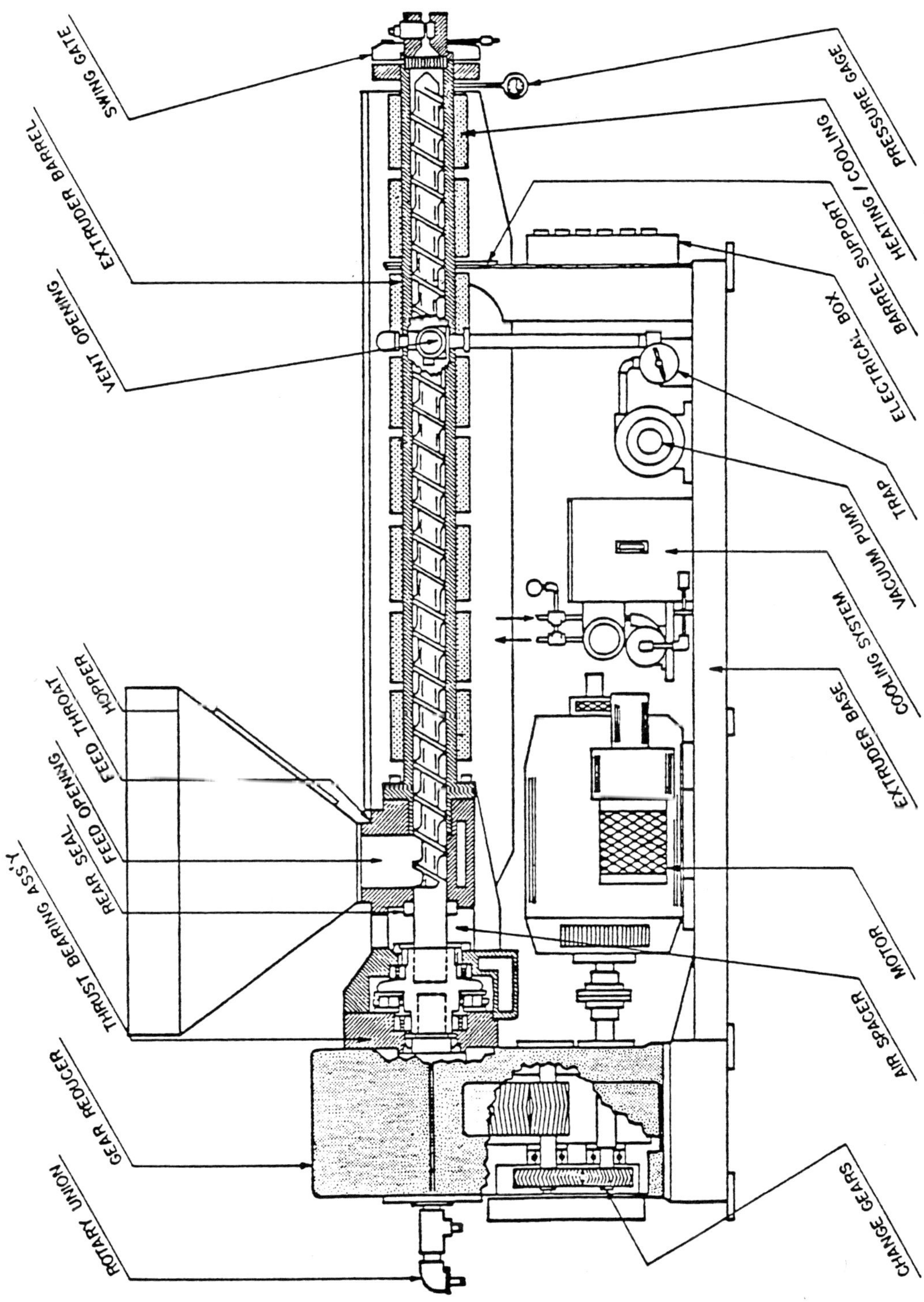

Figure 6-1 A typical single-screw extruder

6.2 THE FUNCTIONS OF AN EXTRUDER

The main functions performed in an extruder are:

1. Conveying
2. Heating and melting
3. Mixing
4. Die forming
5. Degassing (in a vented extruder)

6.2.1 CONVEYING

The forward conveying is due to drag forces between the plastic and the barrel. The drag forces between the plastic and the screw actually work against the forward conveying. Without a drag force against the barrel, there is no forward conveying only rotation with the screw. For good, steady conveying it is important, therefore, to have a large drag force at the barrel and a small drag force against the screw. This is why grooves in the barrel surface at the feed end of the extruder can dramatically improve the conveying characteristics of an extruder. The conveying in the extruder barrel is due to the rotation of the screw. The flow resulting from screw rotation is called drag flow. The conveying through the extruder die is due to the pressure developed by the screw. The flow through the die is due to the difference in pressure between the inlet and outlet of the die. Since the outlet pressure is constant at atmospheric pressure, the extruder output is determined directly by the die inlet or diehead pressure. Thus, the screw is the pressure generating element, while the die is the pressure consuming element. The extruder output, thus, is determined by the interaction of the extruder and the die.

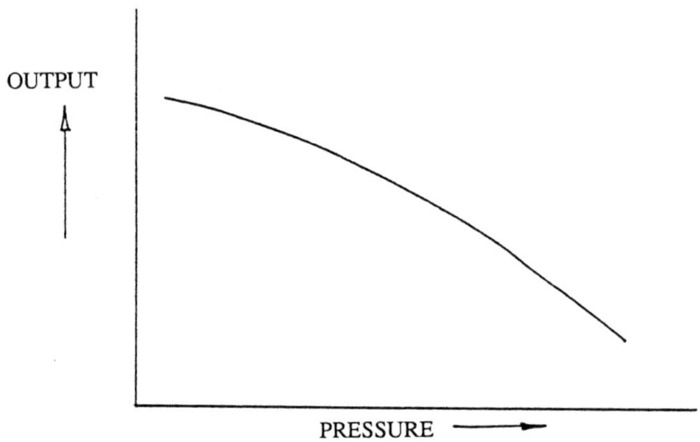

Figure 6-2 Extruder output vs. head pressure

The output of the extruder decreases as the pressure at the end of the screw increases. This is shown in figure 6-2. On the other hand, the output from the die increases as the inlet pressure to the die increases. This is shown in figure 6-3. Thus, each extruder and each die have their own characteristic curves. These are determined by the machine geometry, operating conditions, such as temperature and screw speed, and the flow properties of the plastic. The actual output and head pressure of an extruder/die combination is determined by the intersect of the extruder and die characteristic curves, see figure 6-4.

The die characteristic curve is primarily determined by the geometry of the die flow channel, the die temperature, and the plastic melt viscosity. A die with very thin, restrictive, flow passages will require high pressures to obtain reasonable flow rates, see figure 6-5. A die with large flow passages will require only low pressures to obtain reasonable flow rates, see figure 6-5. Temperature also affects the die characte-

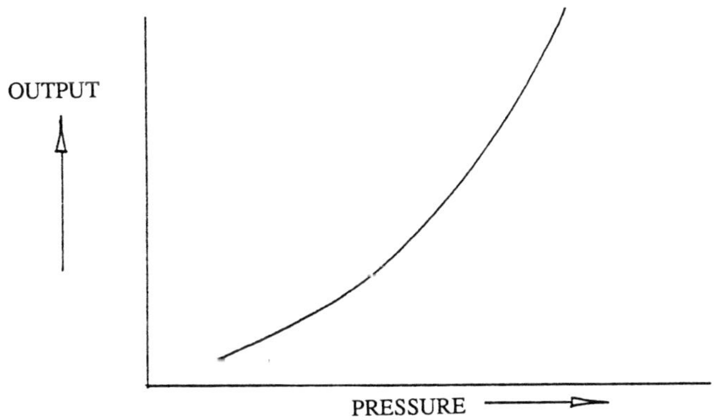

Figure 6-3 Die output vs. die inlet pressure

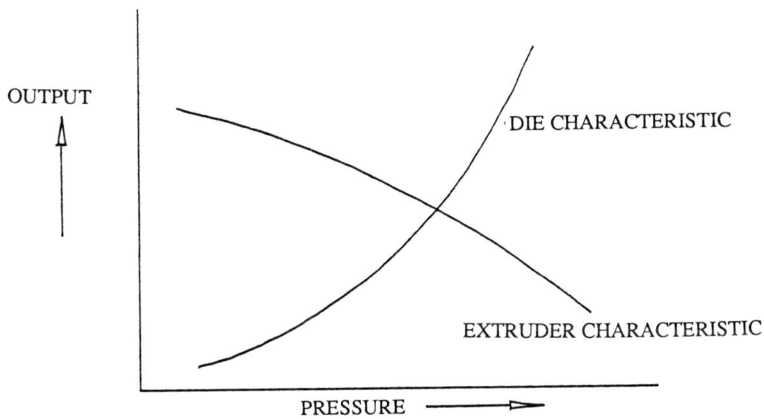

Figure 6-4 Extruder and die characteristics curves

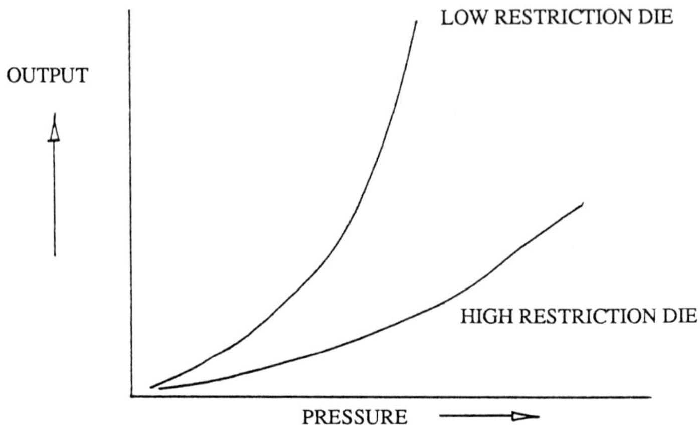

Figure 6-5 Die curves for low and high restriction die

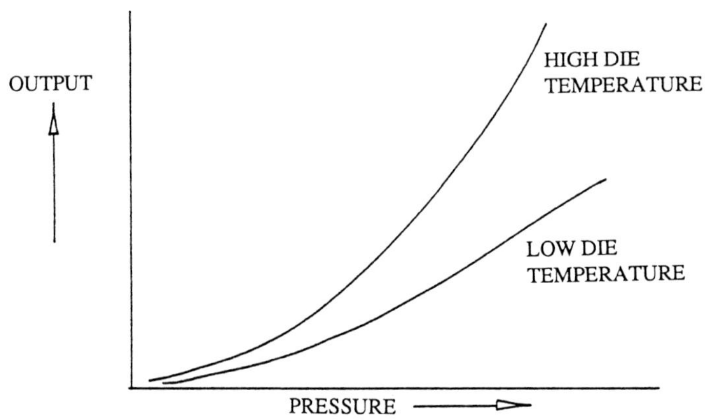

Figure 6-6 Die characteristics for two die temperatures

ristic curve. A high die temperature reduces the restriction to flow and results in a lower diehead pressure at the same output or a higher output at the same diehead pressure. This is shown in figure 6-6. The lower die flow restriction at higher temperature is due to the fact that the plastic melt viscosity reduces with increasing temperature.

The extruder characteristic curve is primarily determined by screw geometry, the screw speed, barrel temperatures, and the flow properties of the plastic. A screw with deep channels will give high output at low head pressure, but the output will drop quickly with pressure. A screw with shallow channels will give low output at low head pressure, but the output will not drop quickly with pressure, see figure 6-7. Therefore, a high restriction die should be combined with a shallow screw, while a low restriction die should be combined with a deep screw, see figure 6-8.

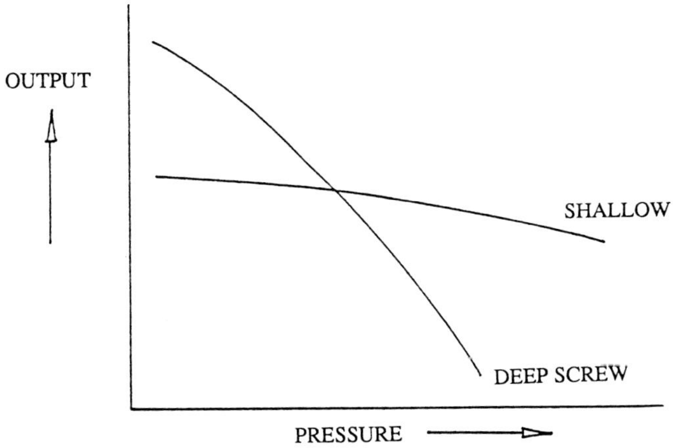

Figure 6-7 Characteristics for two extruder screws

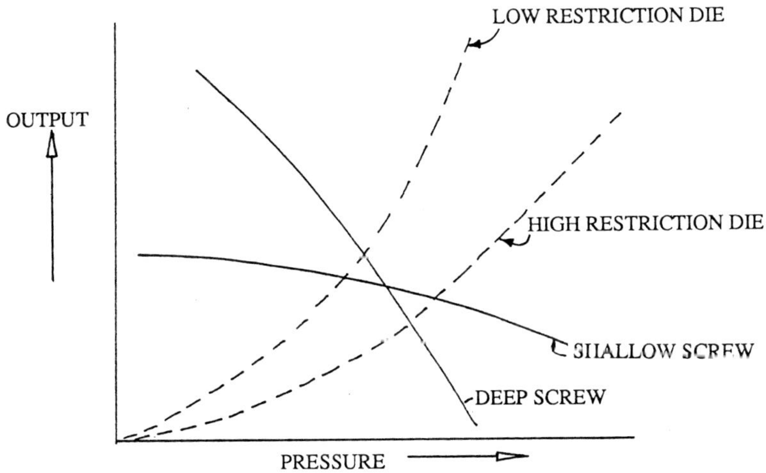

Figure 6-8 Various die and screw characteristic curves

6.2.2 HEATING AND MELTING

There are two sources of heat in the extrusion process. One is the heat from heating elements placed along the extruder, the other is the frictional and viscous heat generation within the plastic. Frictional heating occurs when solid plastic particles slide past a metal surface or past one another. Viscous heating occurs when the plastic melt is sheared by the rotation of the screw. The viscous heating is primarily determined by the shear rate in the melt and the viscosity of the melt. The shear rate is a measure of the shearing action in a fluid. It is determined by the difference in velocity between two points divided by the distance between those points, see figure 6-9.

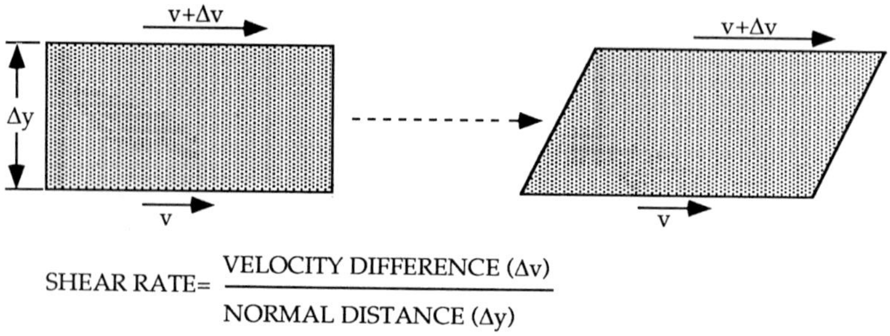

Figure 6-9 Illustration of shear rate

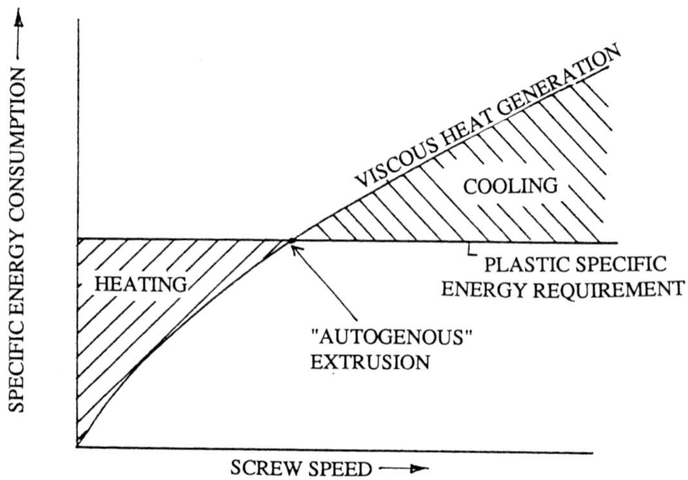

Figure 6-10 Change in viscous heat generation with screw speed

In an extruder the approximate shear rate is the circumferential velocity of the screw divided by the channel depth.

As an example, lets consider a 150 mm (6 inch) extruder running at 90 rpm and a screw channel depth of 7 mm (0.28 inch). The circumference of the screw is 3.14×150 = 471.24 mm. The circumferential speed is 471.24×1.5 = 706.86 mm/sec. The 1.5 comes from 90 rev/min = 1.5 rev/sec. The approximate shear rate becomes 706.86/7 = 100.98 (1/sec). The shear rate is expressed in 1/sec or sec^{-1}; this is called reciprocal seconds. Typical shear rates in a screw channel range from 50 to 100 (sec^{-1}). The shear rates in a extruder increase with screw speed. Because the viscosity of plastic melt is usually quite high, the viscous heat generation is often substantial. In fact, in many extruders most of the heat comes from frictional and viscous heat generation. This is particularly true at high screw speed. At low screw speed the contribution from the barrel heaters is large and the viscous heat generation relatively low. However, as the screw speed increases the contribution from viscous heating increases and the contribution from the barrel heaters reduces. This is illustrated in figure 6-10. The barrel heaters usually contribute only a small amount of heat to the extrusion process. The frictional and viscous heat generation is essentially a transformation of mechanical energy from the extruder drive into thermal energy to increase the plastic temperature.

When the plastic temperature reaches its melting point, melting will start. Melting usually starts at the barrel surface about five diameters from the feed opening. Initially, a melt film forms on the barrel surface. As the melt film grows thicker, a melt pool forms at the leading flank of the flight, pushing the solid bed against the trailing flank of the flight. This model, first described by Maddock (13), is shown in figure 6-11. The two main sources of heat for melting are the heat from the barrel heaters and the viscous heat generation in the melt film between the solid bed and the barrel. For a high melting efficiency, it is important to keep the melt film as thin as possible. The melt film thickness is strongly determined by the screw flight clearance. Therefore, the flight clearance should be kept small. If the flight clearance increases, for instance as a result of screw wear, the melting rate can reduce substantially.

When the barrel temperature is increased, the amount of heat for melting from the barrel heaters increases. At the same time, however, the viscous heat generation in

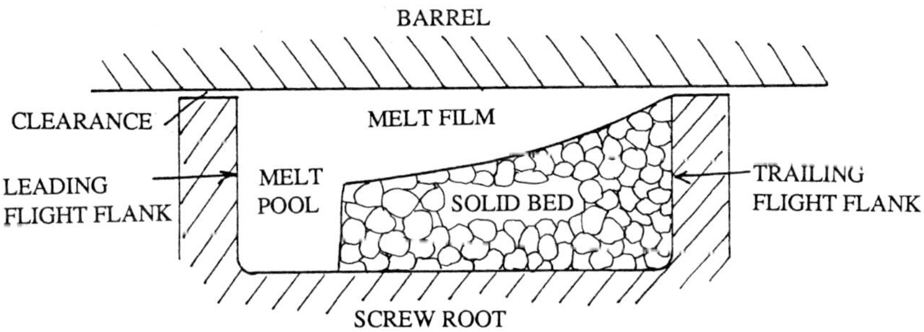

Figure 6-11 Melting model

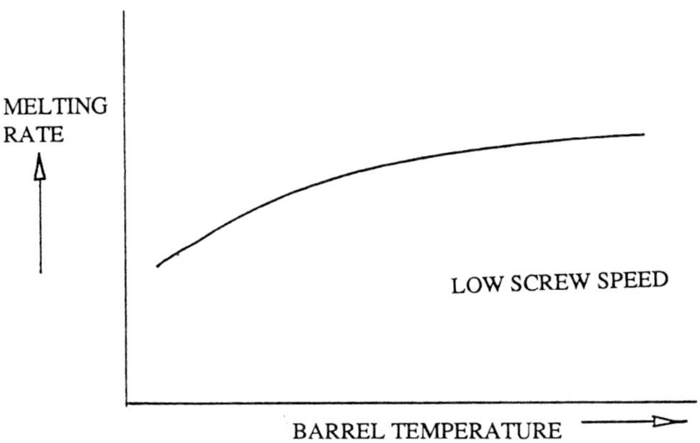

Figure 6-12 Melting rate vs barrel temperature at low screw speed

the melt film will reduce because a higher barrel temperature will reduce the viscosity in the melt film. At low speed, when a major portion of the heat is supplied by the barrel heaters, higher barrel temperatures will increase melting, see figure 6-12. At high screw speed, however, the major portion of the heat for melting comes from viscous heat generation. At high speed, therefore, increased barrel temperature can actually reduce melting in the extruder, see figure 6-13.

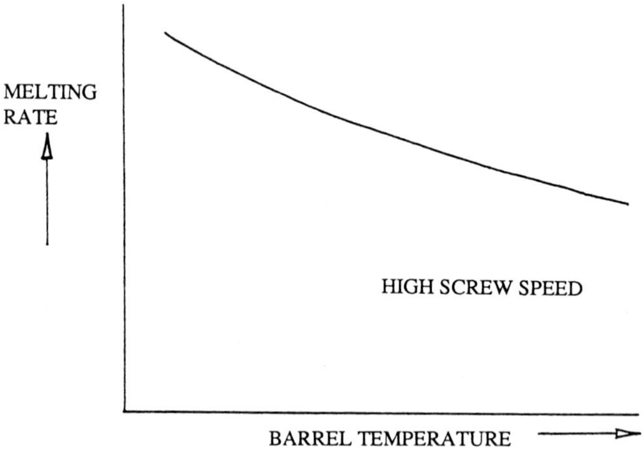

Figure 6-13 Melting rate vs barrel temperature at high screw speed

6.2.3 MIXING

Extruder screws without special mixing devices have relatively poor mixing capability. There are several reasons for this. Material that melts early will be exposed to a long shearing action and will be reasonably well mixed. However, material that melts late will be sheared only for a short time and will not be well mixed. Also, because of the recirculating flow across the screw channel, demixing effects occur in

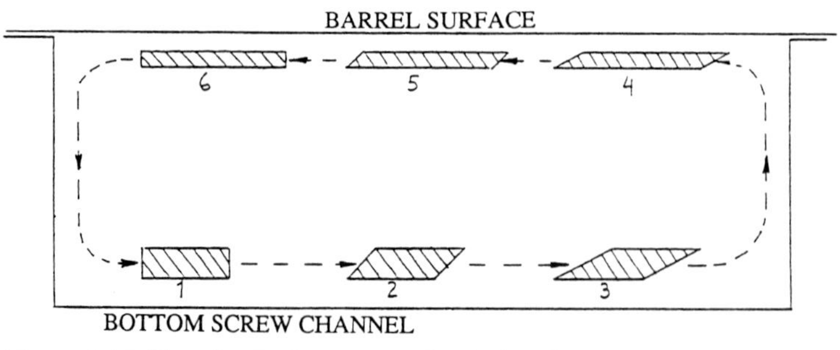

Figure 6-14 Shearing in the outer recirculating flow

the screw channel. This is due to the fact that the direction of shearing in the bottom section of the screw channel is opposite to the shearing upper portion of the channel. This is illustrated in figure 6-14.

The recirculating flow also causes non-uniform temperatures, as shown in figure 6-15. Heat transfer occurs primarily at the barrel surface. As a result, the temperature of the material in the outer recirculating flow will stay relatively close to the barrel temperature. The material in the inner recirculating flow, however, will be insulated from the barrel. Therefore, its temperature can rise substantially above the barrel temperature, by as much as 50 degrees C (90 degrees F) or more. The heat transfer depends strongly on the flight clearance. A large flight clearance reduces the recirculating flow and, thus, even higher melt temperatures can occur over a much larger area of the screw channel. This is shown in figure 6-16.

Because of the inherent non-uniform mixing and melt temperatures in a standard extruder screw, it is important to incorporate efficient mixing devices to improve the melt quality. Mixing is usually divided in distributive mixing and dispersive mixing. Distributive mixing is the mixing of the fluids with similar flow properties. Dispersive mixing is the mixing of fluid with a solid filler. The agglomerates of the filler

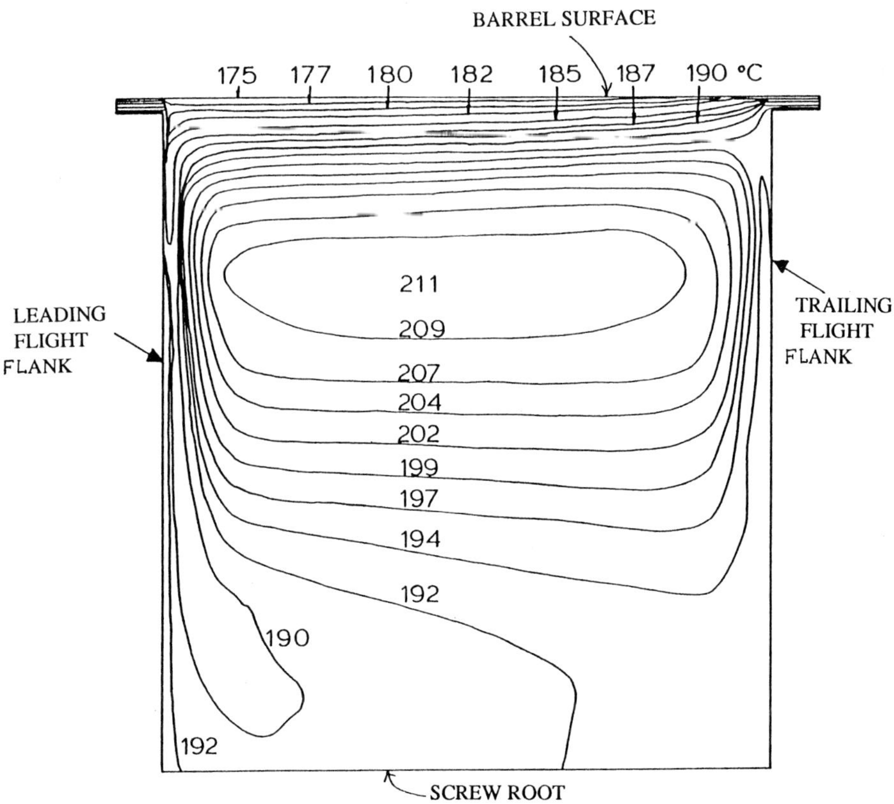

Figure 6-15 Melt temperatures in the screw channel with small flight clearance

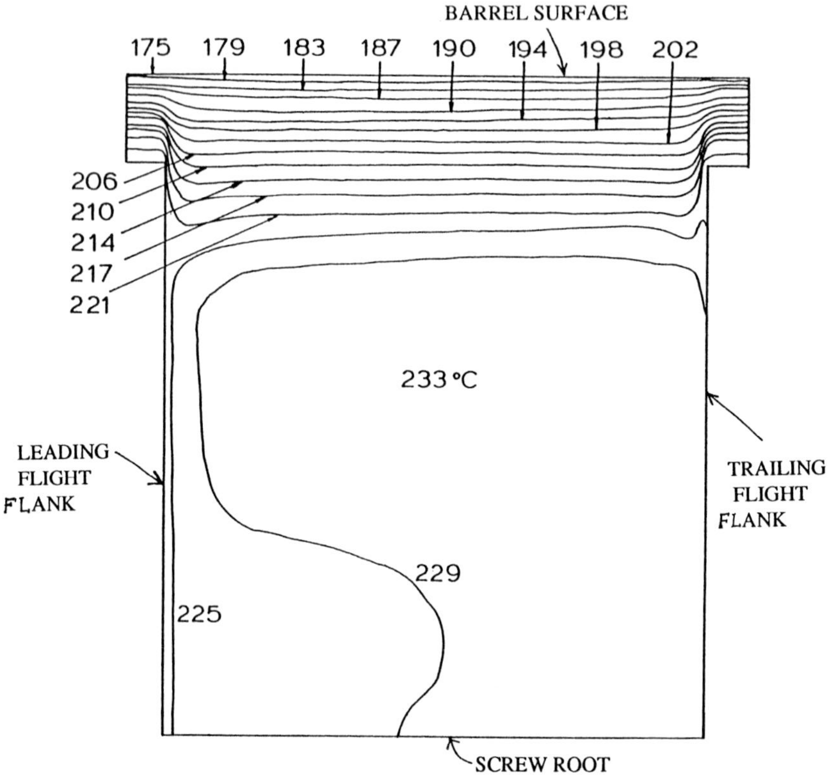

Figure 6-16 Melt temperatures in the screw channel with large flight clearance

have to be broken down by the stresses that the fluid exerts on the filler. Requirements for efficient distributive mixing are:

- large shear strain
- frequent flow splitting
- frequent reorientation

Shear strain is the product of shear rate and shear exposure time. Thus, a shear rate of 100 (1/sec) for 100 seconds will give a shear strain of 10,000. A lower shear rate of 10 (1/sec) will require a longer shear exposure time of 1,000 seconds to achieve the same shear strain. A large shear strain by itself does not necessarily give good mixing. Splitting and reorientation are important in increasing the mixing efficiency substantially.

6.2.4 DIE FORMING

A wide variety of products can be made by extrusion: tubing, pipe, bags, netting, fiber, film, sheet, bottles, fuel tanks, window profiles, etc. For products with a constant cross section, the shape of the extruded product is primarily determined by the geometry of the flow channel in the die. For other products, considerable shaping can occur beyond the die. Examples are corrugated tubing, blow molded products such as bottles, thermoformed products such as pickup truck bed liners, etc. In fact, injection molding can be considered a cyclic extrusion process. Proper design of the extrusion die is very important.

As discussed in section 6.2.1, the flow through the die results from the pressure difference between the inlet and outlet of the die. A number of difficulties can occur in the flow through dies. The restriction to flow can vary along different streamlines in the die. This will result in velocity differences and distortion of the extruded product. Excessive stresses may occur in the die; this can lead to flow instabilities. An example is melt fracture; this causes a severe distortion of the extruded product, see figure 6-17. When the plastic melt leaves the die, it tends to swell. Unfortunately, this swelling is usually non-uniform and distorts the shape of the product. The take up speed is usually higher than the extrusion speed to maintain some tension in the line. This causes a drawdown of the extruded product and changes its size and shape, often resulting in an undesired product geometry. If the extruder screw does not deliver a uniform plastic melt to the die, uneven flow can occur in the die resulting in varying product dimensions. The melt flowing into the die should be uniform in temperature and consistency. Therefore, the extruder screw should have good mixing capability. Uneven melt temperatures can also result in uneven stresses in the product. This can lead to distortion of the product later when the stresses relax.

In hollow products, such as tubing and pipe, the plastic melt has to flow around a central core or mandrel. The mandrel is usually held in place by a number of spider supports, see figure 6-18. As the plastic melt flows past the spider supports it splits and flows together again after the spider. When the melt stream recombines, it takes a certain amount of time before full, intimate contact is established. This is because the long plastic molecules take some time to re-entangle. If insufficient time is avail-

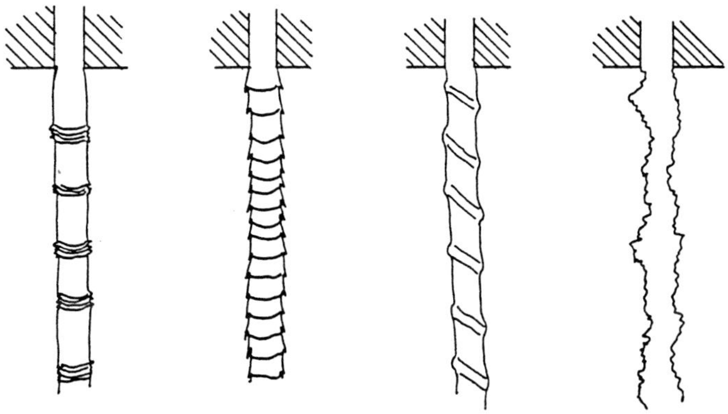

Figure 6-17 Different forms of melt fracture

able for the molecules to re-entangle, a weak spot will form along the length of the extruded product. This is called a weld line or knit line. Such a weld line can cause premature failure of a product, particularly if the product is internally pressurized (which is often the case, e.g. in tubing and pipe). A design for annular products that minimizes the weld line problems is the spiral mandrel die, see figure 6-19.

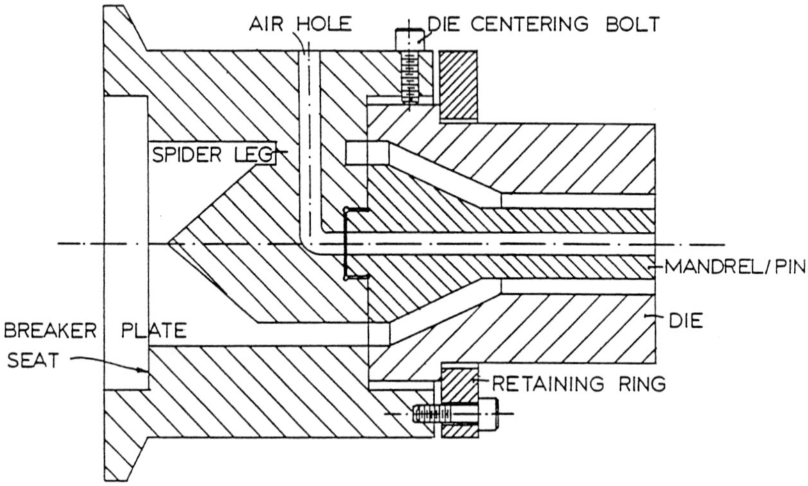

Figure 6-18 In-line tubing or pipe die

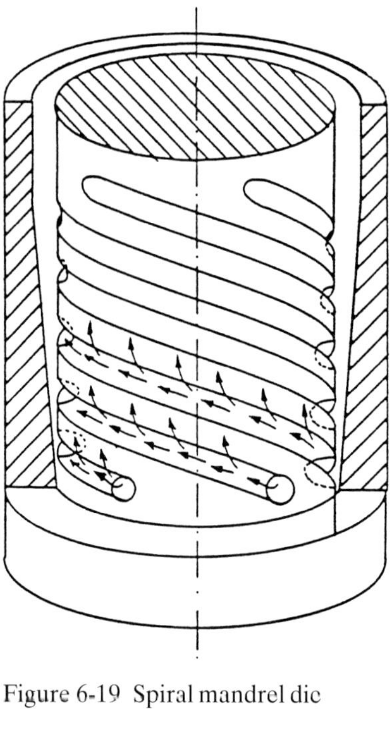

Figure 6-19 Spiral mandrel die

GUIDELINES FOR SHAPE

In order to make a product easy to extrude, the following guidelines should be used regarding the shape of the product.

- Use generous radii at corners (larger than 1 mm).
- Use uniform thickness for exterior walls.
- Avoid very thick walls (larger than 10 mm).
- Minimize the use of hollow sections.
- Make interior walls thinner (about 30%) than exterior walls.
- Make products symmetrical if possible.

Some examples of these guidelines are shown in figure 6-20.

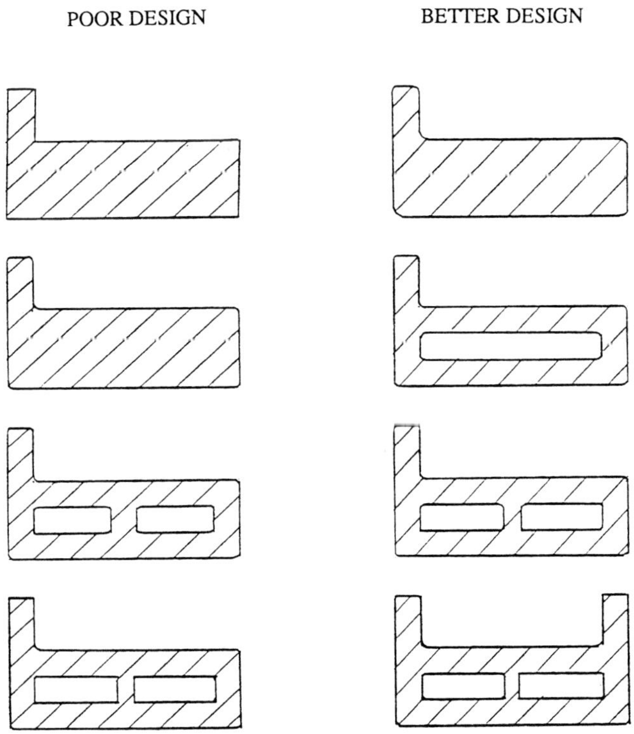

Figure 6-20 Application of die design rules

GUIDELINES FOR DIE DESIGN

The following guidelines can be useful in selecting a die:

- The flow channel geometry should allow streamlined flow without abrupt transitions or dead spots.

- The cross sectional area of the flow channel should reduce gradually from the inlet to outlet to achieve a gradual increase in flow velocity.
- The die should be made to be easy to assemble, disassemble, and clean.
- Spider supports should be located as far from the die exit as possible to minimize weld line problems.
- The land length is usually taken about 10 times the die gap, although larger and smaller values can work well too.
- The total flow restriction along each streamline should be the same to achieve uniform exit velocities. An example is the coathanger die used in flat film and sheet extrusion, see figure 6-21. The length of the preland reduces from the middle to the side to compensate for the longer flow path away from the middle.
- In cases where the flow restriction needs to be adjusted to fine tune the extruded product two methods can be used: mechanical and thermal. The mechanical adjustment involves a change in the flow channel geometry by a moveable part, such as a choker bar, flexlip, die plate, etc. An example of two of those adjustments in a sheet die are shown in figure 6-22. The thermal adjustment involves making a lo-

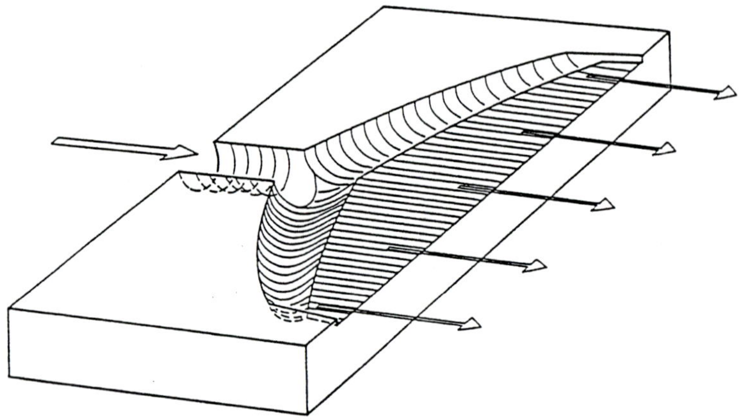

Figure 6-21 Flow distribution in a coathanger sheet die

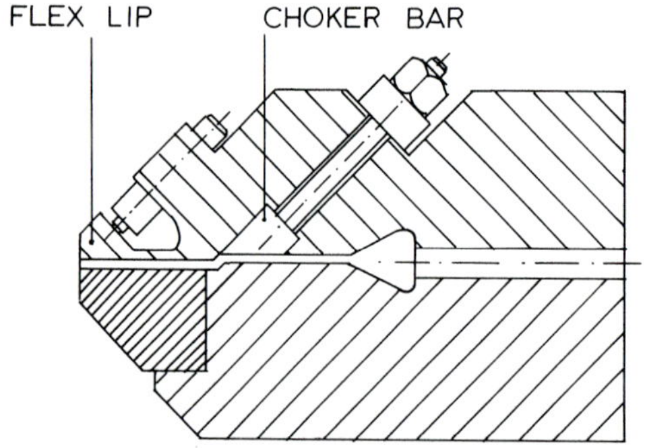

Figure 6-22 Two common sheet die adjustments

cal change in the die temperature to increase or decrease the flow at that point. Some extrusion lines use automatic adjustment of flow restriction based on downstream measurement of the extruded product.

6.2.5 DEVOLATILIZATION

In some extrusion operations a vent opening in the barrel is used to remove volatiles from the plastic. This technique is often used with hygroscopic plastics to lower the moisture level to a level where it does not cause problems in the extruded product. Other volatiles can be removed too, such as monomers, solvents, air, etc. In many cases, removing volatiles through venting is more cost effective than a drying/separation operation before extrusion.

A number of conditions have to be fulfilled to achieve efficient devolatilization:

- The plastic has to be completely melted by the time it reaches the extraction section under the vent port. Thus, the vent port cannot be too close to the feed port.
- The pressure under the vent port has to be zero; therefore, the extraction section of the screw has to have large conveying capability and be only partially filled.
- The screw section just before the extraction section has to be fully filled with plastic melt. This is achieved by incorporating a screw section with low conveying capability.
- The screw section downstream of the extraction section has to have larger conveying capability than the screw section upstream of the extraction section. This can be achieved by increasing the channel depth and/or pitch.

6.3 EFFICIENT EXTRUSION

In the next two sections, we will discuss some of the critical aspects of machine design and process operation for efficient extrusion.

6.3.1 EFFICIENT MACHINE DESIGN

The components of an extruder were listed in section 6.1. We will now list some of the most important aspects of these various components.

6.3.1.1 THE EXTRUDER SCREW

The most important task of the extruder screw is to deliver a well mixed plastic melt to the die at the required pressure and at a constant rate. The following is a listing of guidelines for good screw design:

- Use streamlined design, avoid dead spots.
- The entire barrel surface should be closely wiped by the screw to obtain good heat transfer, mixing, and a narrow distribution of residence times.
- The radial flight clearance between the screw and the barrel should be less than 0.003 times the screw diameter.
- A distributive mixing section should be incorporated to reduce consistency variations and melt temperature non-uniformities. The preferred location of the distributive mixing section is at the very end of the screw.
- A dispersive mixing element should be incorporated when the plastic contains solid filler particles that require high shear stresses to be broken down. Dispersive mixing elements can also be effective in making sure that unmolten plastic particles cannot travel to the end of the screw. Thus, dispersive mixing elements can be useful even if there are no solid filler particles in the plastic.
- Any mixing section used, should have minimum pressure drop and preferably forward pumping capability.
- The screw, barrel, and die should be made from a corrosion resistant material when extruding PVC, fluoropolymers, or other plastics that expose the metal surfaces to corrosive attack.
- When the plastic contains abrasive fillers, such as titanium dioxide, glass, etc., the screw and barrel should be made out of a wear resistant material.
- When a screw coating is used, the preferred coating is one that has low friction properties. This will improve the conveying capability of the screw, resulting in higher output and better stability, and also the ease of cleaning the screw.
- The use of a large radius at the flight flank and screw root is recommended for improved conveying and reduced hangup of material.
- Avoid abrupt changes in the channel depth along the length of the screw. An exception can be made in a multi-stage extruder screw where the transition sections just before and after the extraction section can be made quite short.
- When the extruder is equipped with a grooved barrel section, the screw must have a low compression ratio. The feed section should be shallow and the metering section relatively deep. Also, good mixing capability is critical in grooved barrel extruders.
- When a vent port is incorporated in the barrel, the length of the extruder should be increased to maintain reasonable melting and pressure generating capability.
- When the plastic melt viscosity is high or when the screw discharge pressure is low, the metering section can be made relatively deep. This also applies to situations where a gear pump is placed at the end of the extruder. Conversely, when the viscosity is low or the pressure is high, a shallow metering section should be used.

- When the bulk density of the feed material is low, the depth of the feed section should be made large (20% of the screw diameter or more).

6.3.1.2 THE EXTRUDER BARREL

The extruder barrel has to contain the plastic and fit closely around the screw. The following characteristics are important for trouble free operation:

- The barrel should be straight.
- The barrel should be designed to easily withstand the pressures that occur in the extruder. These can be as high as 70 to 140 MPa (10,000 to 20,000 psi).
- The barrel liner should be more wear resistant than the screw. The screw is easier to replace and less expensive to rebuild than the barrel.
- The downstream barrel support should be a sliding support allowing the barrel to expand when heated. A rigid support can cause warping of the barrel with the possibility of severe damage to screw and barrel.
- For improved solids conveying the feed section of the barrel can be grooved. Good cooling capability should be provided at the grooved barrel section to carry away high frictional heat and to prevent melt from accumulating in the grooves. Also, the grooves should taper gradually to zero depth to minimize the chance of material hangup in the grooves see figure 6-23.
- The vent port, if incorporated, should be offset and tangential to minimize the chance of plastic melt hanging up at the leading edge of the vent port opening. This is shown in figure 6-24. The more conventional symmetrical vent port design tends to give continuous problems with material building up at the bottom of the vent port.

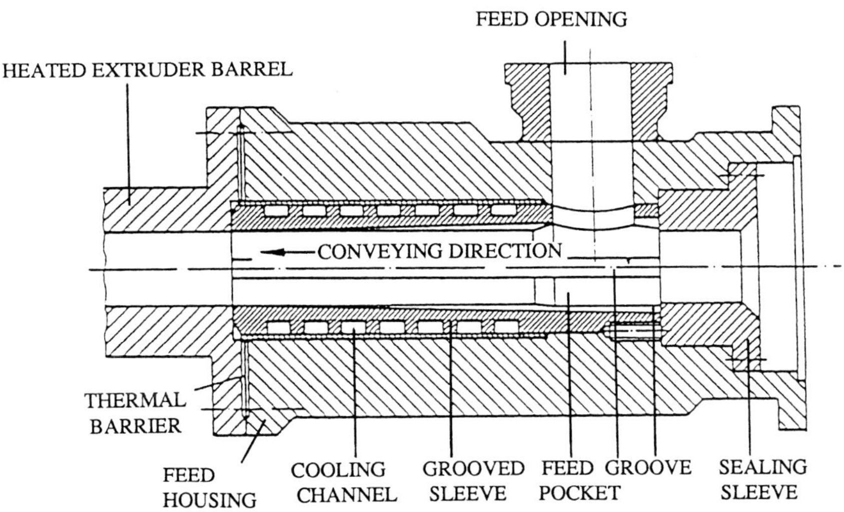

Figure 6-23 Grooved feed section

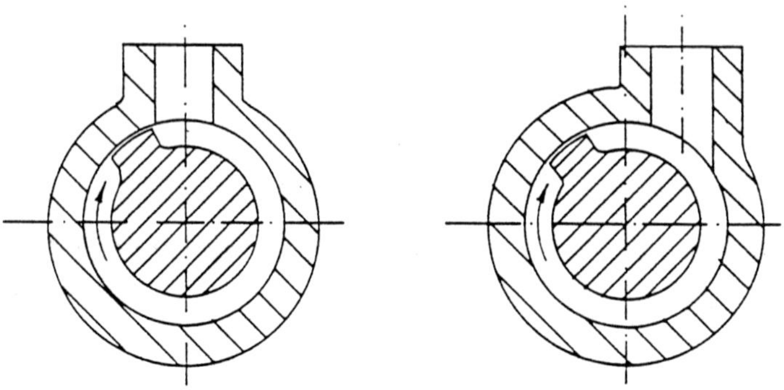

Figure 6-24 Two vent port geometries

6.3.1.3 THE FEED HOPPER AND THROAT

The feed hopper should be designed to give steady flow of the bulk material from the hopper to the extruder. The hopper should have smooth and gradual transitions and avoid sharp corners. The following characteristics are beneficial for stable flow:

- The side walls of the hopper should be steep to reduce the chance of bridging and piping, see figure 6-25.
- The cross sectional shape should be circular not square or rectangular, see figure 6-25. A circular geometry will minimize stagnation.

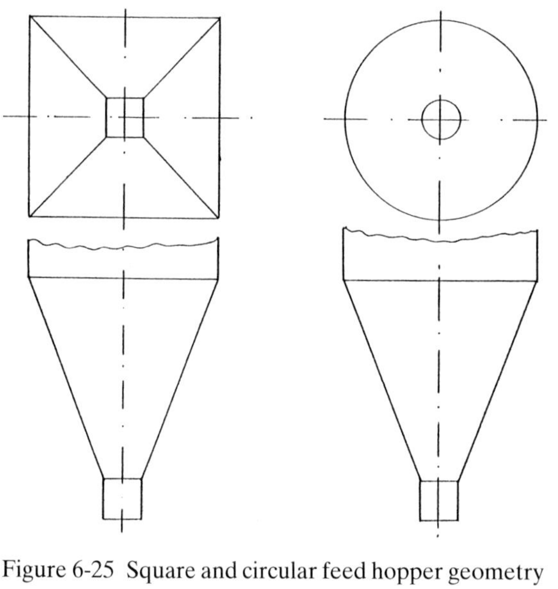

Figure 6-25 Square and circular feed hopper geometry

- A low friction coating on the inside hopper surface reduces hangup of material.
- For difficult bulk materials special features can be used to promote steady flow, examples are vibrating pads and crammer feed, see figure 6-26.

The feed throat should be designed to provide smooth flow from the feed hopper to the screw channel. The following characteristics are important:

- The feed throat should have good cooling capability to avoid premature heating of the plastic. This can cause plastic sticking to the walls, resulting in lower output and unstable flow.
- At the barrel side of the feed throat should be a thermal barrier to avoid excessive heat flow from the barrel to the feed throat, see figure 6-27.
- The axial length of the feed opening should be considerably larger than the screw pitch to avoid flight induced flow instabilities, see figure 6-28.
- The width of the feed opening should be smaller than the barrel diameter to maintain good forward flow, see figure 6-28.
- The feed opening should be offset, preferably tangential, to maximize the intake capability of the screw, see figure 6-28.

It should be noted that on many commercial extruders the geometry of the feed hopper and throat differs considerably from the desired geometry described above. This

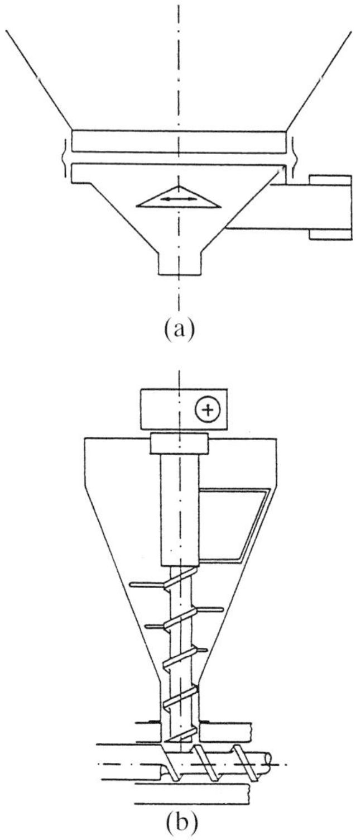

Figure 6-26 (a) Bin Activator (b) Crammer feeder with stripper arm

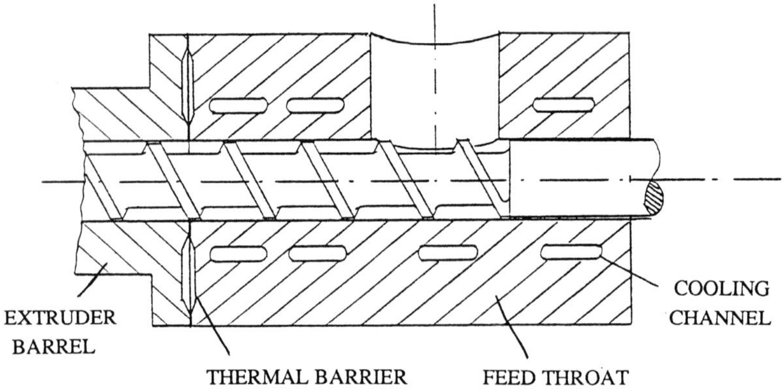

Figure 6-27 Feed throat with thermal barrier

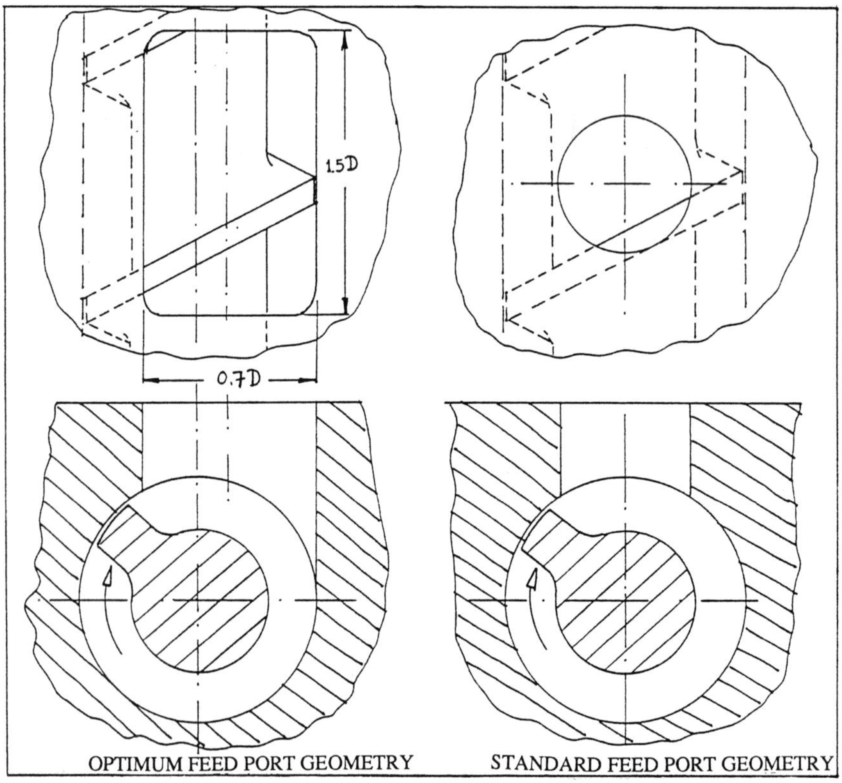

Figure 6-28 Different feed port geometries

partially results from the fact that sometimes machinery manufacturers put more emphasis on ease of manufacture than on functional performance. For processors, however, functional performance should be first priority in machine selection.

6.3.1.4 THE EXTRUDER DRIVE

For the drive the following characteristics are important for efficient extrusion:

- The drive should have good speed regulation. Digital drives provide speed regulation of 0.01% full speed or better. Good speed regulation is particularly important when the extruder is operated at relatively low speed.
- The motor should have a direct coupling with the reducer. Many extruders use a belt transmission between the motor and reducer. However, these create a considerable power loss and cause slippage with resulting screw speed variation.
- An overtorque or overcurrent shutoff should be provided to prevent damage to the extruder.

6.3.1.5 INSTRUMENTATION AND CONTROL

The minimum requirements for instrumentation were discussed in chapter 5. Some important considerations are:

- The barrel temperature should be measured and controlled close to the inside barrel surface; thus, a deep well temperature sensor should be used.
- Dual sensor temperature control can result in reduced temperature fluctuations.
- On/off temperature control should be avoided in extrusion, whether used for heating or cooling.
- Melt temperature sensors should be in contact with the melt and designed to minimize conduction errors. Many combined pressure/temperature transducers do not give a good indication of the melt temperature.
- When PID (proportional/integral/derivative) control is used, the controller has to be carefully tuned under actual process conditions. Improper tuning can result in very poor temperature control. Some controllers set their tuning parameter automatically; these are called auto-tuning controllers.
- Capillary type pressure transducers filled with mercury cannot be used in extruded products used in food packaging or medical products.
- Piezoresistive pressure transducers using silicon-on-sapphire technology can improve dynamic response, measurement accuracy and precision, and susceptibility to damage.
- A data acquisition system, DAS, is very useful in capturing data from the process and allowing careful analysis. Some systems can process the data and present statistical information such as control charts, standard deviation, process capability, alarms for presence of assignable causes, etc. Such capability is very useful in implementing SPC.

6.3.2 EFFICIENT PROCESS OPERATION

An efficient extrusion process requires not only a good machine but also good operation of the process. Some important aspects of efficient extruder operation will be discussed next.

6.3.2.1. FEED STOCK CONSISTENCY

It is impossible to have a stable process without having a consistent feed stock. At this point in time, the only way that a consistent feed stock quality can be ensured is by using material quality control (QC) to check critical material properties. Unfortunately, one cannot always rely on the resin supplier to deliver a material with consistent properties (e.g. 14). There are a number of benefits to in-house testing. It allows the processor to establish a data base on materials actually used in production. Thus, non-conforming lots can be identified before problems occur on the production floor. Also, the material variability from different resin suppliers can be compared quantitatively. This can provide useful selection criteria. Another benefit is that a company that practices materials QC is less likely to receive out of specification material (assuming the resin supplier knows that the material will be tested). Unfortunately, the opposite is true as well. A company that is not practicing materials QC is more likely to receive out of specification material. Without materials QC it can be very difficult and costly to prove that a problem is caused by inconsistent feed stock properties.

One of the most commonly used material properties is the melt index (MI), also called melt flow index (MFI). Unfortunately, MI is one of the least useful properties because:

A) The measurement error can be very large (±20% or more), particularly for fractional MI plastics.
B) The MI is not a good indicator of the extrusion characteristics of a plastic.

A major problem with the MI measurement is that only very large variations in the resin can be positively identified. Smaller changes cannot because of the large measurement error. As a result, the MI measurement is not very discriminating. Unfortunately, the MI test is very popular with resin suppliers even though it is not very useful for processors.

A better indicator of extrusion characteristics is the viscosity as a function shear rate as determined on a rheometer, such as a capillary rheometer or cone-and-plate rheometer. Such data, however, are often not available from resin suppliers even though they are much more useful and discriminating than MI. A property that is very important in extrusion operations where the material is stretched, is the elongational viscosity. Examples of such extrusion operations are blow molding, fiber spinning, biaxially oriented film, blown film, etc. The elongational viscosity is usually not measured even though simple tests are available to analyze the stretching behavior of plastic melts (e.g. 15).

One of the important questions in materials QC is what properties to measure. Often properties are measured that may be important for end product performance but not for processability, examples are environmental stress crack resistance, impact properties, flammability, etc. Properties that are important in extrusion are:

- bulk density
- compressibility
- internal coefficient of friction
- external coefficient of friction
- melting point
- induction time at extrusion temperature
- stabilizer level and type
- viscosity vs. strain rate and temperature
- specific heat vs. temperature
- thermal conductivity/diffusivity
- level of other additives (filler, lubricants, etc.)

Clearly, it is not always practical to measure all the properties important in extrusion. However, with a basic rheometer and a differential scanning calorimeter many important properties can be readily determined. Bulk density and compressibility can be easily determined with a scale, a container, and a piston with weights. Thus, good testing of processing related characteristics does not require an extraordinary investment in instruments or manpower. Most instruments, nowadays, are programmable with computer controlled data acquisition capability. This means that the manpower necessary to run these instruments is minimal.

6.3.2.2 TEMPERATURES

Typical process temperatures for semi-crystalline plastics are about 50 degrees C above the melting point. Amorphous plastics are usually processed about 100 degrees C above the glass transition temperature. On the one hand, we would like to keep the process temperatures high to reduce the melt viscosity and make the material more easily processable. On the other hand, we would like to keep the temperatures low to reduce the chance of degradation. As a result, a thermally stable plastic may have a wide operating window. However, a thermally sensitive plastic may have only a narrow temperature range within which it can be successfully processed.

The temperature profile along the extruder can have a strong effect on the extruder performance. The following guidelines can be used to set the temperature profile:

- Set the barrel temperature in the feed section such that the diehead pressure fluctuation is minimum. This requires a graphical display of the pressure as a function

of time, either using a chart recorder or a CRT. Alternatively, the barrel temperature in the feed section can be set such that the motor load is maximum.
- Set the transition section to the temperature where the melt temperature variation is minimum.
- Set the metering section and die temperature at the value at which the melt temperature should be controlled.

Some extruders have a melt temperature feed back control. This typically changes the set point of the last one or two barrel temperature zones in response to a change in melt temperature. Such a system can only react to very slow changes in melt temperature. It should be noted also that reduced barrel temperature in the metering section will increase the pumping capacity of the metering section. Also, increased screw temperature increases the metering section pumping capacity.

6.3.2.3 SCREEN PACK

In many extrusion operations a screen pack is used to filter contaminants so that they do not end up in the final product. A drawback of the screen pack is that its flow restriction increases as contamination builds up. This will cause a reduction in output. Unless the take-up is reduced correspondingly, this will change the product dimensions. One way to approach this is to use a pressure feedback control that changes the screw speed in such a way as to keep the pressure behind (downstream) the screen pack constant.
Another approach is to change the screen pack before the pressure drop across the screens changes too much. This may not work too well because it can lead to frequent screen changes (increased down time and scrap). Also, as we have seen in chapter 5, small changes in pressure can cause relatively large changes in output.

A better approach would be to use an automatic screen changer that allows a rapid screen change, preferably without having to stop the line. Yet another possibility is to use a gear pump to control the output. With a gear pump the screw speed is usually varied in such a way as to maintain constant inlet pressure to the gear pump.

6.3.2.4 FEEDING

Most single screw extruders are flood fed. Thus, the screw takes in as much material as it can. In this case, the screw is essentially completely filled with plastic and the output is determined by the screw speed. Another approach is to use starve feeding. In starve feeding the feed material is metered into the feed throat at a rate below the flood feed rate. Thus, the material does not buildup in the feed hopper. As a result, this is one way to avoid flow problems in the feed hopper. The extruder is only partially filled in starve feeding; the degree of fill is dependent on the feed rate and the screw speed. A higher screw speed reduces the degree of fill, a higher output increases the degree of fill. In starve feeding the screw speed no longer determines the output. By changing feed rate or screw speed the effective L/D (length to diameter ratio) of the extruder can be varied. A number of advantages can be obtained from starve feeding:

- The motor load and screw torque can be reduced.
- The specific energy consumption, SEC, can be varied.

The SEC determines how much mixing and viscous heat generation occurs in the plastic. Starve feeding, thus, allows an additional degree of process control. It can be very useful in cases where excessive motor load and/or excessive melt temperatures occur.

Another feeding technique that can be quite useful is gravimetric output control, GOC. In GOC, small amounts of feed material are weighed and discharged with a measurement of the time elapsed between each discharge. This is typically about 15 to 30 seconds. Thus, the mass flow rate is measured about every 15 to 30 seconds. If the mass flow rate changes either the screw or the line speed can be adjusted to maintain the same mass per length of extruded product. GOC has found widespread use, particularly in multi-layer coextrusion where it is used to control the average thickness of each individual layer. GOC usually allows an output control of 0.5 percent, which is quite good. In a way, GOC can be seen as an alternative to a gear pump, although it is quite a different system. GOC can be used in certain applications where a gear pump might present a serious problem. For example, when the plastic contains a high level of abrasive filler or when the plastic has limited thermal stability.

6.3.2.5 GEAR PUMPS

Gear pumps consist of two closely intermeshing gears that fit closely in the housing, see figure 6-29. The gear pump is, in essence, a counterrotating twin screw extruder with very short screws and very large flight pitch. Gear pumps offer a number of benefits:

- Good output stability, output variation less than 1.0 percent.
- High volumetric efficiency, usually better than 95 percent.
- Good pressure generating capability.

On the other hand, there are some disadvantages as well:

- Poor mixing capability.
- Low energy efficiency, typically 20 to 50 percent.
- Susceptible to wear.
- Expensive.

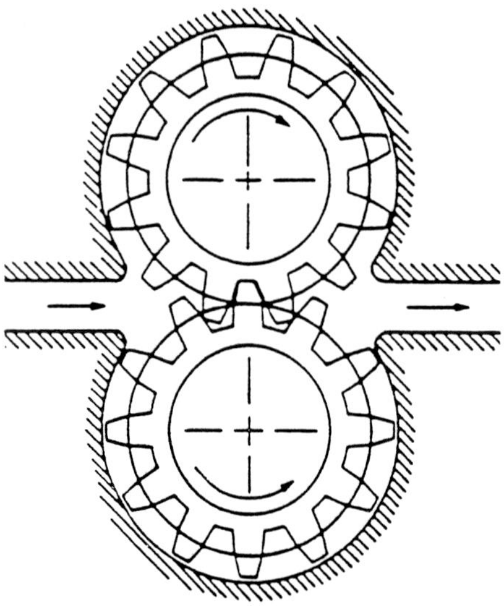

Figure 6-29 Gear pump

Good applications for gear pumps are:

1. To generate pressure for extruders that have low pressure generating capability, such as non-intermeshing twin screw extruders, corotating twin screw extruders, vented single screw extruders.
2. To maintain good output uniformity in processes where the product dimensions have to be tightly controlled. Examples are fiber spinning and medical tubing. A regular extruder usually can keep output fluctuations as low as 1 percent. A gear pump can keep output fluctuations to less than 1 percent, from about 0.2 to 0.5 percent in the best case.

The following situations have a substantial chance of problems when a gear pump is used:

1. When the plastic contains a high level of abrasive filler.
2. When the plastic has poor thermal stability.

The reason that degradation can easily occur in gear pumps is that not all the plastic melt flows in the spaces between the teeth where it is conveyed efficiently. A small amount of melt flows to the sides of the gears where it is used to lubricate the gears. This lubrication flow can cause a small amount of plastic to remain in the gear pump for a long time. The long residence time combined with the high temperature in the gear pump will make degradation likely to occur.

7 ADVANCED STATISTICAL TOOLS

In chapters 1 through 5 some basic concepts of statistical process control were introduced. In some cases it may be helpful to use more advanced statistical tools for a more detailed process analysis. Some useful advanced statistical techniques are analysis of variance, design of experiments, regression analysis, response surface methodology, and the Taguchi method. These subjects will be discussed next.

7.1 ANALYSIS OF VARIANCE

Analysis of variance (ANOVA) is used to separate the contribution of several factors to the total variation observed in a variable. As discussed in chapter 1, the variance (v) is the sum of the squared deviations (SS) divided by the sample size n minus one. The sample size minus one is equal to the degrees of freedom (DF). The variance is equal to the standard deviation squared (s^2). Thus, the variance can be expressed as:

$$v = s^2 = \frac{SS}{DF} = \frac{\Sigma(x_i - \bar{x})^2}{n-1} = \frac{\Sigma(x_i)^2 - n\bar{x}^2}{n-1} \tag{7-1}$$

where $\bar{x}$ is the sample mean.

We will illustrate the use of ANOVA by analyzing extruded sheet thickness. The sheet is about eight inch wide and has a nominal thickness of 0.010 inch. The thick-

TABLE 7-1. THICKNESS DATA OF EXTRUDED SHEET

	1	2	3	4	5	6	7	8	total	mean
1	10.2	10.4	9.8	9.7	10.4	10.5	10.3	9.9	81.2	10.1500
2	10.2	10.3	9.7	9.7	10.3	10.4	10.3	9.8	80.7	10.0875
3	10.3	10.3	9.8	9.7	10.4	10.5	10.4	9.8	81.2	10.1500
4	10.2	10.2	9.7	9.6	10.4	10.5	10.4	9.9	80.9	10.1125
5	10.2	10.3	9.7	9.7	10.5	10.6	10.3	9.8	81.1	10.1375
6	10.3	10.3	9.7	9.7	10.4	10.6	10.3	9.8	81.1	10.1375
7	10.3	10.3	9.8	9.8	10.4	10.6	10.2	9.9	81.3	10.1625
8	10.2	10.4	9.8	9.8	10.3	10.5	10.3	9.9	81.2	10.1500
9	10.2	10.3	9.8	9.7	10.4	10.6	10.3	10.0	81.3	10.1625
10	10.2	10.3	9.7	9.7	10.4	10.6	10.3	9.9	81.1	10.1375
11	10.2	10.3	9.7	9.6	10.5	10.5	10.2	9.9	80.9	10.1125
total	112.5	113.4	107.2	106.7	114.4	115.9	113.3	108.6	892.0	
mean	10.227	10.309	9.745	9.70	10.4	10.536	10.3	9.873	10.127	10.1364

ness is measured in eight evenly spaced locations across the sheet. These measurements are repeated every three feet for a length of thirty feet. The thickness measurements are shown in table 7-1, the thickness is expressed in one-thousandth inch.

7.1.1 ONE-WAY ANOVA

We will first do a one-way analysis of variance to determine whether the difference in the columns is significant. This corresponds to the differences in thickness across the width of the sheet. To perform the ANOVA we have to calculate various sums of squares. The total sums of squares, SST, can be calculated as follows:

$$SST = s^2\,(DF) = \Sigma x_{ij}^2 - N\bar{\bar{x}}^2 \qquad (7\text{-}2)$$

where $\bar{\bar{x}}$ is the grand average and N the number of data.
From the data in the table we determine the value SST=8.064. The sums of squares attributable to differences in position (columns) can be calculated from the averages for the eight columns:

$$SSC = r\Sigma\bar{x}_j^2 - N\bar{\bar{x}}^2 \qquad (7\text{-}3)$$

where $\bar{x}_j$ is the average for column j and r is the number of rows.
With the expression above we obtain SSC=7.574. The column sums of squares, SSC, is the squared deviation accounted for by the differences between columns (position). The deviation within columns is unexplained. In this one-way ANOVA the deviation within columns is considered a residual. Thus, the residual, SSR, is determined from:

$$SSR=SST-SSC \qquad (7\text{-}4)$$

We can now construct a ANOVA table.

TABLE 7-2. TABLE FOR ONE-WAY ANOVA

Source	SS	DF	MS	F	p
Total	8.064	88-1=87	–	–	–
Position	7.574	8-1=7	1.082	176.653	<<0.01
Residual	0.490	87-7=80	0.006125	–	–

The mean squared deviation, MS, is determined by dividing the sums of squares by the degrees of freedom, DF. For instance, MS-position is SS-position (7.574) divided by the DF-position (7), this makes MS-position=1.082. The F ratio is determined by the ratio between the column MS and the residual MS. The F value represents the ratio between the explained and the unexplained variances. If the actual F value is larger than the F value from the F-table the factor is significant; the level of the significance is indicated by the probability value, p.

The value of F at a certain significance is determined from the F-table in appendix I. This table gives F values significant at 1% or 0.01. The DF in the numerator for the position is 7, which makes n_1=7. The DF in the denominator is 80, which makes n_2=80. There is no actual column for n_1=7, but there is one for n_1=6 and for n_1=8. Thus, we have to find the F value for n_1=8 by interpolation. The same situation occurs for the row. There is no row for n_2=80, however, there is one for n_2=60 and n_2=120. We can again determine the F value for n_2=80 by interpolation. This leads to a value of F≈2.9. Clearly, the F value for position is much larger than the F value from the F-table; therefore, the position is a significant factor for the thickness variation of the extruded sheet.

7.1.2 TWO-WAY ANOVA

If we look at the actual data in table 7-1, we see that there is thickness variation not only between columns, indicating a variation across the extrusion direction, but also within individual columns. This indicates a variation in extrusion direction, which could be due to a surging of the extruder output. To analyze both the variation in extrusion direction and across the extrusion direction, we can perform a two-way analysis of variance.

For this analysis we require also the sums of squares for the rows. This is determined from:

$$SSR = c\Sigma\bar{x}_i^2 - N\bar{\bar{x}}^2 \qquad (7\text{-}5)$$

where c is the number of columns and $\bar{x}_i$ the row averages.
With the actual data in table 7-1 we find for the sums of squares for the rows SSR=0.0436. We can now construct a two-way ANOVA table.

TABLE 7-3. TABLE FOR TWO-WAY ANOVA

Source	SS	DF	MS	F	p
Total	8.064	88-1-87	–	–	–
Position	7.574	8-1=7	1.082	169.86	<<0.01
Columns				F(.01)=2.9	
Length	0.044	11-1=10	0.0044	0.69	>0.01
Rows				F(.01)=2.6	
Residual	0.446	87-7-10=70	0.00637	–	–

The two-way ANOVA shows that the position across the extrusion direction is a significant source of variation, while the variation in extrusion direction is not significant. Significant thickness variation across the extrusion direction indicates that the geometry of the flow channel of the die is inappropriate or that there are consistency differences across the flow stream. Inappropriate flow channel geometry may be due to improper adjustment of the choker bar and/or flex lip. If the adjustment of the choker bar and/or flex lip does not significantly reduce the thickness variation, the basic geometry of the die may be incorrect.

7.1.3 THE BASIC IDEAS BEHIND ANOVA

ANOVA leads to the determination of a probability that the variation associated with a particular factor is significant; let us call this factor 1. The probability is based on the ratio of the mean squared deviation of the factor 1, MS1, to the mean squared deviation of the residual, MSR. The ratio MS1/MSR is the F ratio, thus, F is a ratio of variances. F is used to determine the probability that two variances belong to the same population. If this probability is small, then we conclude that they are not from the same population.

Each variance associated with a factor is a measure of the relative effect of the factor on the dependent variable being analyzed; in our example the dependent variable is the sheet thickness. Knowing the relative effect is not sufficient. We need to know how important the variance is compared to some objective standard. The objective standard used is the natural variation of the dependent variable in steady state. This is assessed by the mean squared deviation of the residual. Also, we need to know if other factors contribute to the variability.

The natural variation in a process is caused by small, random changes in control variables. This variation is considered "noise" and may or may not be measured. This noise is superimposed on the variation associated with the various factors being analyzed. When many random variables contribute to noise, the sum of their contributions will tend toward zero, however, they will contribute some variance. Thus, the F ratio between the variance of a factor and the noise variance indicates whether or not the factor is significant. A large F ratio indicates that the factor is significant, large is considered a value greater than 1. When F is small, this indicates that the variation is largely due to the system noise.

The level of probability associated with F determines the statistical significance. When the steady-state noise level is very low, factors that are actually weak may show a high statistical significance. It is important, therefore, to distinguish between statistical and practical significance. Statistical significance indicates simply that the variation associated with a factor can be distinguished from the system noise. In some cases where very precise measurements can be made or with very large samples, it is may be possible to detect very slight variations that may not have a practical significance. If the variation has no practical significance it should be ignored.

Another practical problem is that steady-state noise cannot always be separated from unsteady-state noise. In the design of an experiment, we should always try to separate the effects of unsteady behavior; however, this is not always possible. If it cannot be done the residual variance will be composed of both steady and unsteady-state noise. There are two sources for the residual variance. One is short-and long-term variation due to variables such as relative humidity, ambient temperature, power surges, instrument drift, raw material changes, etc. The other is the inherent measurement error at a particular time.

For a good estimate of the system noise we need repeated measurements spaced as close as possible in time. With repeated measurements we can quantify the measurement error and, thus, obtain a better estimate of the system noise. With an independent estimate of the system noise the F test can be used to determine whether the remaining variance, i.e. the residual, is significant. If it is, this means that we have not identified at least one factor affecting the system.

7.2 DESIGN OF EXPERIMENTS

7.2.1 INTRODUCTION

One of the most useful statistical tools in extrusion process engineering is the statistical design of experiments, DOE. This is used to systematically study the effect of certain variables on a particular process or product characteristic. In DOE jargon, the variables are called factors and the characteristic that may be affected by these factors is termed the response. For instance, if we wanted to study the effect of screw speed and barrel temperature on extruder output, the factors would be screw speed and barrel temperature and the response would be the extruder output.

Certain terms used in DOE are different from terminology normally used in extrusion experiments or in the polymer field in general. This has to do with the fact that the original concepts of DOE were developed in agricultural research. Some of the important terms will be defined first. A **"factor"** is a variable whose effect on a **"response"** is investigated. The factor may or may not be controllable. The **"levels"** of a factor are the values of that factor that are used in the experiment. The levels could be different screw speeds, different extruder screws, different drying times of the compound, etc.

A **"treatment"** is a particular level used for a factor. A **"treatment combination"** is the set of levels used for the various factors in a particular experimental run. For instance, this could be a screw speed of 40 rpm, a 400°F barrel temperature, and a low intensity mixing screw. The **"experimental environment"** consists of the surrounding conditions that can influence the result of the experiment. A **"block"** is a part of the experimental material or of the experimental environment that is likely to be more homogeneous within itself than between other parts. For instance, parts produced within one shift are likely to be more uniform than parts from different shifts. A group of samples produced within one shift would be considered as a block.

In the design of experiments the uniformity within blocks is often used to minimize the effect of unwanted variables. Setting up experiments that utilize the uniformity within blocks is called **"planned grouping;"** experiments designed this way are called **"block designs."** Another technique that is used to eliminate the uncontrolled variables is **"randomization."** When the sequence of experiments occurs in a purely chance manner, the experiment is said to be randomized. This enables the application of tests of significance, the construction of confidence intervals, and improves the validity of estimates of experimental error. A technique to balance out the effect of uncontrolled factors is **"replication."** Replication is the repetition of a measurement or observation in order to increase precision or to provide the means for measuring precision.

Requirements for sound experimentation are:

1. The objectives of the experiment should be clearly defined.
2. The effect of unwanted factors should be minimized.
3. The experiment should be free from bias.
4. The experiment should provide a measure of experimental error variance (precision).
5. The precision of the experiments should be sufficient to meet the objectives of the experiment.

Replication and randomization are important tools in making sure that the experiment meets the requirements listed above. The following flow chart lists the steps involved in planning an experimental program.

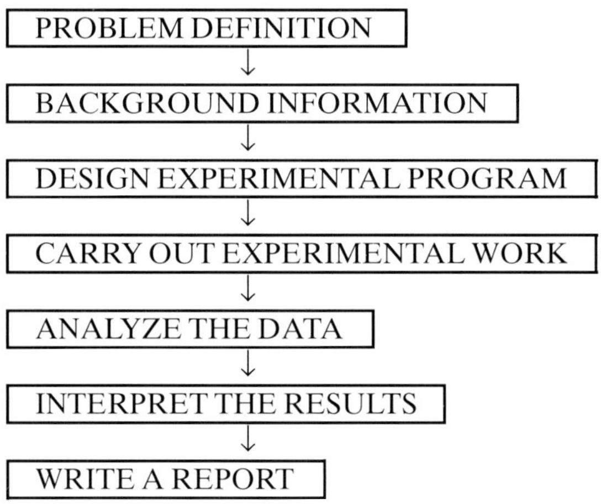

7.2.2 FACTORIAL DESIGNS

Various experimental designs are used, such as completely randomized designs, block designs, Latin square, Youden square, nested designs, etc. For experiments that involve several variables at two or more levels the factorial design is particularly useful. Several types of factorial designs are used, the main ones being the full factorial, the blocked factorial, and the fractional factorial .

In a full factorial design all combinations of variables at every level are run. If we have N levels and K variables, the total number of experimental runs for a full factorial design is N^K. Thus, for a two-level design (N=2) examining the effect of three factors (K=3), or variables, the number of experiments will be eight (2^3). In a full factorial design there are three pieces of information that are sought. The first is an estimate of the effects of various factors, second an estimate of the possible interactions between factors, and third an estimate of the variance.

When the number of experimental runs is too large to be carried out under uniform conditions, a blocked factorial design can be used. In such a design the full set of combinations of factors and levels is divided into subsets so that certain higher order interactions are related to blocks. Each subset constitutes a block and all blocks are run in the experimental program. In a blocked factorial design certain higher order interactions cannot be estimated.

When the total number of experimental runs is too large to run all combinations, a fractional factorial design can be used. The number of runs increases very rapidly with the number of factors, a two-level design with three factors requires eight runs, four factors requires sixteen runs, five factors requires thirty-two runs, etc. When a considerable number of factors is involved, it may not be practical to carry out all experimental runs. In a fractional factorial design only a subset of the full factorial is run. In a fractional factorial design not all interaction effects can be estimated and the design may not provide enough information to estimate the variance.

Efficient screening designs were developed by Plackett and Burman (25). These saturated fractional factorial designs require only N trials to evaluate (N-1) factors, but they do not provide estimates of interactions between variables.

In a factorial design the following information can be obtained:

1. An estimate and comparison of the effects of several factors.
2. An estimate of possible interaction effects.
3. An estimate of the variance.

7.2.3 WHY USE FACTORIAL DESIGNS?

Factorial designs are very useful because they provide the maximum amount of information from the minimum number of experiments. The typical approach in experimental work is to study one variable at a time by varying the level of that variable over a certain range. This makes for a very simple experimental plan, however, this approach does not work well in many situations.

Let us take an example from the extrusion field. Suppose we want to study how the output of a screw melt pump depends on the helix angle of the screw flight and the channel depth of the screw (flight height). All other factors are kept constant. If the polymer melt behaves as a Newtonian fluid, the output as a function of helix angle and channel depth can be calculated from extrusion theory (16). This relationship can be graphically representation by a three-dimensional surface plot, as shown in figure 7-1.

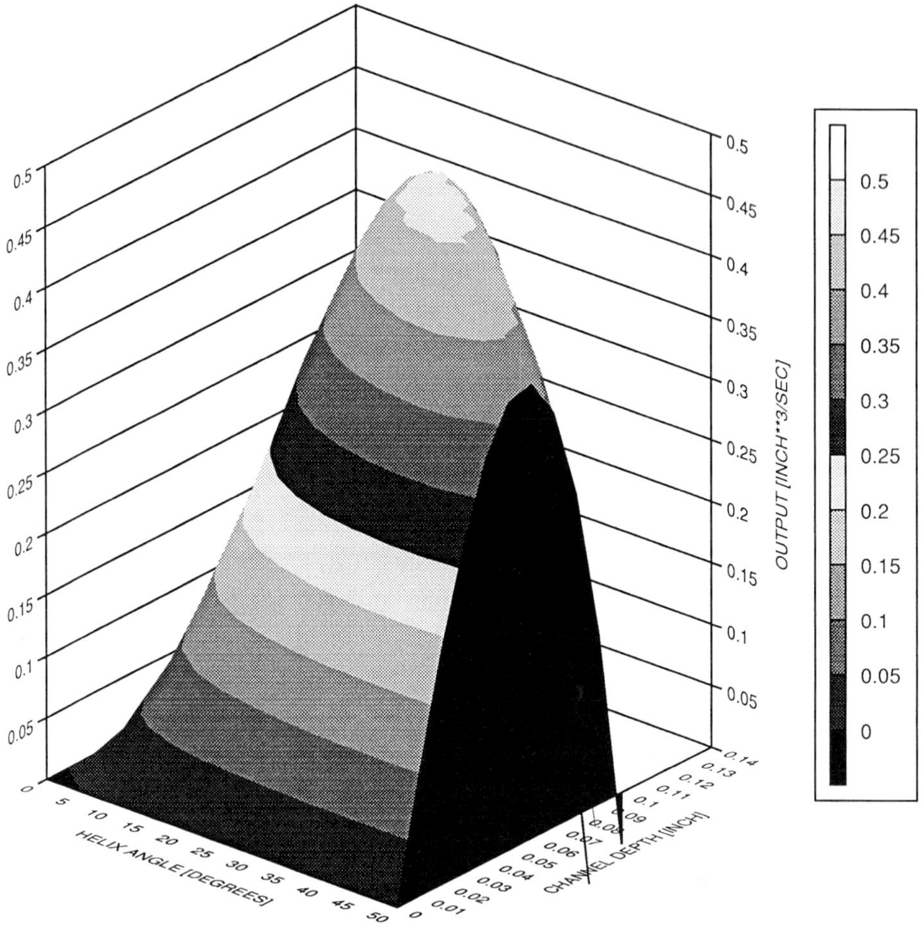

Figure 7-1 Three dimensional surface plot of output vs. helix angle and channel depth

The curved surface in figure 7-1 looks like a steep mountain. If we want to operate at the highest output, we need to find the optimum values of the helix angle and channel depth. Another way of graphically representing this data is through a contour plot, see figure 7-2.

The contour plot shows lines of equal output. It is a convenient method to determine the optimum operating conditions. If we assume that the information shown in figures 7-1 and 7-2 is not known, we can do two simple one-at-a-time experiments to determine the optimum operating point. First, we run trials at three different channel depth values, H=0.08", H=0.10", and H=0.12", all at a constant helix angle of 15 degrees. The corresponding output values are found to be 0.34, 0.38, and 0.40 $inch^3$/sec, respectively, as shown in figures 7-1 and 7-2. Clearly, based on these results it is most advantageous to use a channel depth of 0.12".

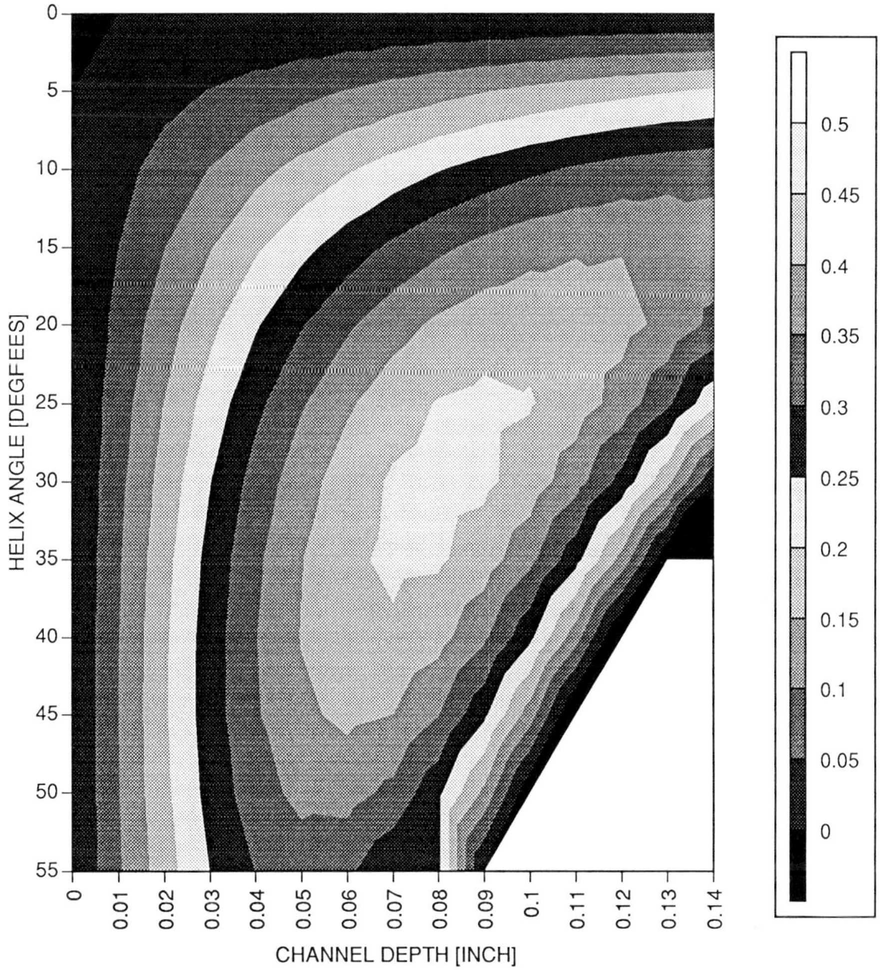

Figure 7-2 Contour plot of output vs. helix angle and channel depth

Next, we run trials at three different helix angles, ø=5°, ø=15°, and ø=25°, all at a channel depth of 0.12". The corresponding output values are determined to be 0.19, 0.40, and 0.38 $inch^3$/sec, respectively. From these results it seems safe to conclude that a helix angle of 15 degrees and a channel depth of 0.12 inch will yield the highest output, its value being 0.40 $inch^3$/sec. If we look at the contour plot, however, it can be seen that the highest output is obtained at a helix angle of 30 degrees and a channel depth of about 0.08 inch. The true maximum output is 0.47 $inch^3$/sec. Thus, the result of the one-at-a-time experiments have lead us to an operating point that is relatively far removed from the true optimum. Instead of running at an output of 0.47 $inch^3$/sec, we only achieve 0.40 $inch^3$/sec.

7.2.4 INTERACTION EFFECTS

This example illustrates why one-at-a-time experiments can lead to sub-optimal results. Therefore, this approach is generally not recommended. The reason that the one-at-a-time experiments lead to the incorrect optimum value in this example has to do with how the factors helix angle and channel depth influence the response. The effect of the helix angle on output changes when the value of the channel depth is changed. In statistical terminology we say that there is interaction between factors helix angle and channel depth. In factorial experiments interactions between factors can be uncovered and taken into account in determining the optimum operating point. Interaction between factors often occurs in process engineering problems and this is why factorial experiments are so much more useful that one-at-a-time experiments.

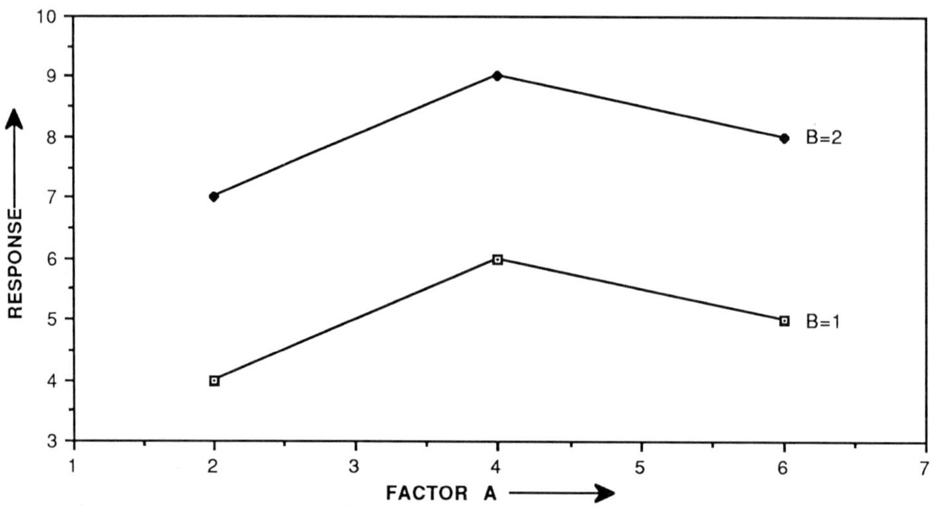

Figure 7-3 Example of two factors without interaction

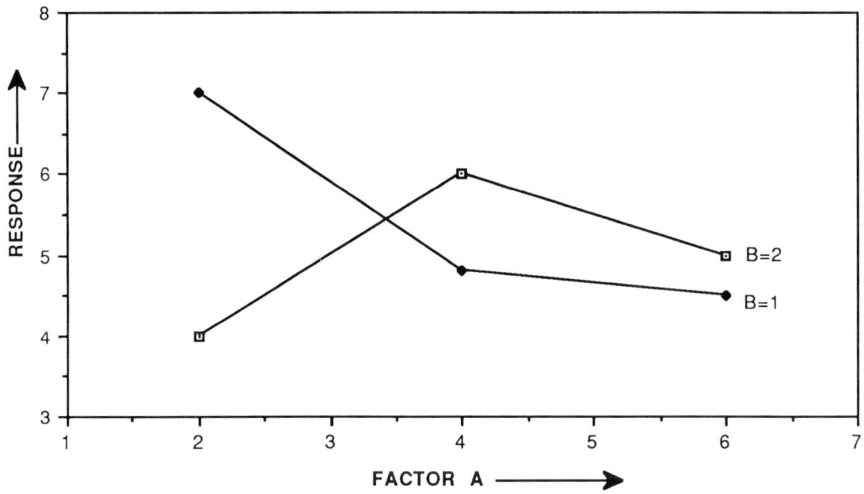

Figure 7-4 Example of interaction between factors A and B

If the difference in response between two levels of factor A remains constant over different levels of factor B, then there is no interaction between factor A and B. In other words the AB interaction is zero. This shows up graphically as a vertical shift of the curves plotting response versus factor A at different values of factor B, see figure 7-3.

If the difference in response between two levels of factor A changes over different levels of factor B, then there is interaction between factors A and B. Thus, the AB interaction has a non-zero value. In a plot of response versus factor A, the curves at different values of factor B will not be parallel, see figure 7-4.

If there are more than two levels of A and B the interaction AB can be composed of more than one component. If there are *a* levels of factor A and *b* levels of factor B, then the interaction has (a-1) (b-1) independent components. Interaction between two factors is called a second order effect. Higher order interaction can occur with three or more factors. For example, the ABC interaction is the interaction between factor C and the AB interaction; it is called a third order effect.

7.2.5 TWO FACTOR FACTORIAL EXPERIMENT

The two factor experiment is the simplest multifactor experiment. We can take an example of a 1.5″ extruder where we measure the melt temperature at the end of the extruder at different screw speeds and barrel temperatures. The data are shown in table 7-4.

TABLE 7-4. DATA ON MELT TEMPERATURE AT VARIOUS SCREW SPEEDS AND BARREL TEMPERATURES

Screw speed	Barrel 175 °C	Barrel 185 °C	Barrel 195 °C	Row total	Row average
20 rpm	185.2	191.3	199.8	$R_1 = 576.3$	192.100
60 rpm	209.1	213.7	219.0	$R_2 = 641.8$	213.933
100 rpm	213.5	217.4	222.1	$R_3 = 653.0$	217.667
140 rpm	216.7	218.6	223.2	$R_4 = 658.5$	219.500
Column total	$C_1 = 824.5$	$C_2 = 841.0$	$C_3 = 864.1$	2529.6	
Col. average	206.125	210.25	216.025		210.8

Analysis of variance can be used to test whether the melt temperature is dependent on the screw speed and/or the barrel temperature. The ANOVA table is constructed as follows:

number of rows: $r = 4$
number of columns: $c = 3$
grand total: $T_g = 2529.6$
total number of observations: $N = r.c = 12$
correction factor: $C = T_g^2/N = 533239.68$
sum of squares of rows: $SSR = \Sigma R_i^2/c - C = 1447.0467$
sum of squares of columns: $SSC = \Sigma C_j^2/r - C = 197.835$
total corrected sum of squares: $SST = \Sigma x^2 - C = 1663.1$
error sum of squares: $SSE = SST\text{-}SSR\text{-}SSC = 18.2183$

The ANOVA table for this two-factor factorial is shown in table 7-5.

TABLE 7-5. ANOVA TABLE FOR TWO-FACTOR FACTORIAL

Source of variation	Sum of squares (SS)	Degrees of freedom (DF)	Mean square MS = SS/DF	F*	F value from table
Rows (screw speed)	1447.047	(r–1) = 3	MSR = 482.35	158.86 = MSR/MSE	9.78**
Columns (barrel temp)	197.835	(c–1) = 2	MSC = 98.92	32.58 = MSC/MSE	10.92**
Error	18.218	(r–1) (c–1) = 6	MSE = 3.0364		
Total	1663.100	(rc–1) = 11			

F* = mean square of source/mean square of error
** = significant at 1 % level

We can conclude that statistically significant effects exists between rows if the F-value for the rows (row mean square/error mean square) is greater than $F_{1\text{-}\alpha}$ for (r–1) and (r–1) (c–1) degrees of freedom. The F-value for 3 and 6 degrees of freedom is 9.78, see F table in appendix I. The F value for the rows is significantly higher than the $F_{1\text{-}\alpha}$ value, therefore, we have to conclude that statistically significant effects exist. The same is true for the columns. Thus, both the screw speed and the barrel temperature have a statistically significant effect on the melt temperature. Obviously, the effect of screw speed is larger than that of barrel temperature as indicated by the large F-value for the rows (158.86) as compared to the F-value for the columns (32.58).

The experiment described above used only one observation for each level of the factors involved, in other words one observation per cell. For a good estimate of the error of variance the observations must be replicated. The analysis of variance of a replicated two-factor crossed factorial design is similar to that of an unreplicated design. The ANOVA table is constructed as follows:

number of rows is r
number of columns is c
number of observations per cell is k
total number of observations is N = krc
grand total is T_g
correction factor $C = T_g^2/N$
sum of squares of rows is $SSR = \Sigma R_i^2/(kc) - C$
sum of squares of columns is $SSC = \Sigma C_j^2/(kr) - C$
interaction sum of squares is $SSI = (cell_{ij}\ total)^2/k\text{-}SSR\text{-}SSC\text{-}C$
total sum of squares is $SST = \Sigma x^2 - C$
error sum of squares is SSE = SST-SSR-SSC-SSI

The ANOVA table for the replicated two-factor factorial is shown in table 7-6.

TABLE 7-6. ANOVA TABLE FOR A REPLICATED TWO-FACTOR FACTORIAL DESIGN

Source of variation	Sum of squares	Degrees of freedom	Mean square	F
Rows	SSR	r–1	MSR = SSR/(r–1)	MSR/MSE
Columns	SSC	c–1	MSC = SSR/(c–1)	MSC/MSE
Interaction	SSI	(r–1) (c–1)	MSI = SSI/(r–1) (c–1)	MSI/MSE
Error	SSE	rc(k–1)	MSE = SSE/rc(k–1)	
Total	SS	krc–1		

7.2.6 TWO-LEVEL FACTORIAL DESIGN WITH K FACTORS

The 2^K factorial designs are frequently used in industrial applications. This design allows separate estimation of the individual effects and the interaction effects of the K factors. Each trial is determined by the combination of levels of the factors with the total number of trials being 2^K. Different notations are used to identify the different trials. The common notation is to label each factor by a letter and to show the two levels of the factor by a plus (+) and a minus (–) sign. For a four-factor factorial the 2^4

trials are shown in table 7-7. The actual trials are shown in the unshaded portion of the table; the shaded portion of the table is used to determine the various interactions.

TABLE 7.7. TABLE OF SIGNS FOR A 2^4 FACTORIAL

A	B	C	D	AB	AC	AD	BC	BD	CD	ABC	ABD	ACD	BCD	ABCD
–	–	–	–	+	+	+	+	+	+	–	–	–	–	+
+	–	–	–	–	–	–	+	+	+	+	+	+	–	–
–	+	–	–	–	+	+	–	–	+	+	–	–	+	–
+	+	–	–	+	–	–	–	–	+	–	+	+	+	+
–	–	+	–	+	–	+	–	+	–	+	–	+	+	–
+	–	+	–	–	+	–	–	+	–	–	+	–	+	+
–	+	+	–	–	–	+	+	–	–	–	–	+		+
+	+	+	–	+	+	–	+	–	–	+	+	–	–	–
–	–	–	+	+	+	–	+	–	–	–	+	+	–	–
+	–	–	+	–	–	+	+	–	–	+	–	–	+	+
–	+	–	+	–	+	–	–	+	–	+	+	+	+	+
+	+	–	+	+	–	+	–	+	–	–	–	–	–	–
–	–	+	+	+	–	–	–	–	+	+	+	–	–	+
+	–	+	+	–	+	+	–	–	+	–	–	+	–	–
–	+	+	+	–	–	–	+	+	+	–	+	–	–	–
+	+	+	+	+	+	+	+	+	+	+	–	+	+	+

In a 2^K factorial design all K first-order effect can be estimated; further, all K(K–1)/2! two-factor interactions, all K(K–1) (K–2)/3! three factor interactions, etc. Thus, in a four factor design we can estimate four first-order effects, six two-factor interactions, four three-factor interactions, and one four-factor interaction for a total of 16 statistics. The magnitude and signs of these statistics are not influenced by one another; the statistics are said to be orthogonal. The determination of the different interactions is shown in table 7-7, each interaction has its own column. The sign of each interaction is found by multiplying the + and – signs of the components of that interaction. For example, the signs of the AB interaction is found by multiplying the sign of the A factor with the sign of the B factor, the sign of the ABC interaction is found by multiplying the sign of the AB interaction with the sign of the C factor, etc.

The estimated interaction is determined by adding all the observations carrying the + sign and taking their average, then all observations carrying a – sign are added and averaged. The estimated interaction is then the difference between these two values. For instance, the estimated AB interaction is:

$$(x_1+x_4+x_5+x_8+x_9+x_{12}+x_{13}+x_{16})/8 - (x_2+x_3+x_6+x_7+x_{10}+x_{11}+x_{14}+x_{15})/8$$

The ABC interaction is:

$$(x_2+x_3+x_5+x_8+x_{10}+x_{11}+x_{13}+x_{16})/8 - (x_1+x_4+x_6+x_7+x_9+x_{12}+x_{14}+x_{15})/8$$

In shorthand notation the interaction is calculated by $\bar{x}_+ - \bar{x}_-$ with the first term representing the average of the plus-values and the second term the average of the minus-values. There is a faster method to compute the interactions, this is the Yates' algorithm (18).

Factorial designs are useful to perform a preliminary analysis of the effect of several variables. The drawback is that it does not yield quantitative information of the effect of a variable. The factorial design can thus be used as a first step to uncover the most important variables, those influencing the process the most. The next step is to use regression analysis on additional experimental trials to develop a quantitative relationship of the effect of the different variables. Obviously, it is not necessary that the design of experiments and regression analysis are done in separate steps. The two processes can be combined and can also include response surface analysis. In the experimental trials for regression analysis it is important to examine the variables at several levels so that more exact relationships can be determined.

7.3 REGRESSION ANALYSIS

Regression analysis is used to find a functional relationship between one or more independent variables and a dependent variable. First, we have to determine whether or not two variables are dependent. This can be done using correlation analysis.

7.3.1 CORRELATION ANALYSIS

Correlation analysis involves calculating a correlation coefficient (r), which is a measure of the dependence between the two variables. Then we test the hypothesis that the observed r could be observed by chance if the two variables were actually independent. This technique can be applied to normally distributed data. Non-normal data can sometimes be transformed into normal ones by using logx, x^2, 1/x, etc.

For variables x and y with n paired samples the correlation coefficient is calculated as follows:

$$r = \frac{\Sigma(x_i - \bar{x})(y_i - \bar{y})}{\left[\Sigma(x_i - \bar{x})^2 \Sigma(y_i - \bar{y})^2\right]^{0.5}} \tag{7-6}$$

The variance r is basically the variance of x and y together divided by the product of the individual variances. An alternate expression for the variance is:

$$r = \frac{\Sigma x_i y_i - N\overline{xy}}{\left[(\Sigma x_i^2 - N\bar{x}^2)(\Sigma y_i^2 - N\bar{y}^2)\right]^{0.5}} \tag{7-7}$$

The value of r can range from –1 to +1. A value of –1 indicates a perfect negative correlation; a value of +1 indicates a perfect positive correlation; a value of 0 indicates no correlation. When $r = 0$ the variables are independent.

In real life situations there will be chance variations and r probably will not be zero even if the variables are independent. The statistical analysis proceeds as follows, first we assume that the variables are independent and then we determine the probability of observing a greater r by chance alone. If the probability is below some threshold value, usually 1 % or 5 %, we reject the hypothesis and conclude that the apparent relationship is not due to chance alone. The analysis does not tell us, however, which variable is the dependent one and which the independent one. This information has to be found some other way. The r values whose probability of being observed in samples of various sizes are 0.01, 0.02, 0.05, and 0.10 are shown in the r table in appendix II, for more complete tables see reference 21.

As an example, suppose we have 12 pairs of x and y data and the calculated r value is $r = 0.72$. In the table we look at the row for 10 degrees of freedom (12-2) and find that the probability of observing an r greater than 0.708 is 0.01 or 1 %. With the actual $r = 0.72$ we can be more than 99 % confident that $r \neq 0$ and x and y are dependent.

The square of the correlation coefficient, r^2, is called the coefficient of determination. It is indication of the fraction of variation in y that can be accounted for by variations in x and vice versa. However, the correlation coefficient does not tell us how one variable changes when the other is changed. For this type of information we need to use regression analysis.

7.3.2 LINEAR REGRESSION

Linear regression is used when a change in one variable causes a proportional change in another variable. Such a relationship can be mathematically described by a first order polynomial; higher order polynomials will be discussed under nonlinear, polynomial regression. The equation for a linear or proportional relationship is:

$$y = b_0 + b_1 x \tag{7-8}$$

In this equation b_0 is the intercept of the line with the y-axis and b_1 is the slope of the line. Linear regression can sometimes be used on variables that have a nonlinear relationship by transforming the data. This can be done using logx, sinx, x^2, 1/x, etc.

An example in polymer processing is the relationship between polymer melt viscosity and shear rate. For many polymers this relationship is highly nonlinear. However, by using the log transformation for both the viscosity and shear rate, the relationship becomes very close to linear in many cases and linear regression can be used. This has led to the commonly used power law equation relating melt viscosity to shear rate. The appropriate transformation can often be determined by plotting the data in various forms. For instance, when melt viscosity is plotted against shear rate on a log-log graph, the log transformation becomes immediately obvious.

The objective in regression analysis is to determine the equation that best fits the data. The criterion for best fit is the least squared error. This means that the sum of the squared deviations between the points and the line is minimized. The squaring is done because we want to minimize the absolute value of the distance between the actual data points and the line. In other words, we want the points to be as close to the line as possible whether they are above or below the line. The best fit slope can be determined from the following equation:

$$b_1 = \frac{N\Sigma x_i y_i - \Sigma x_i \Sigma y_i}{N\Sigma x_i^2 - (\Sigma x_i)^2} \tag{7-9}$$

where N is the number of datapoints (pairs of x,y data).
The best fit intercept can be determined from:

$$b_0 = \frac{\Sigma y_i}{N} - b_1 \frac{\Sigma x_i}{N} = \bar{y} - b_1 \bar{x} \tag{7-10}$$

The equation for b_1 is similar to the one for r and they are in fact related. The probability of observing b_1 by chance alone is the same as that of observing the corresponding r by chance alone. The coefficient of determination, r^2, is a measure of how well the line fits the data; when $r^2 = 1$ there is a perfect fit and when $r^2 = 0$ there is no fit at all. The difference is that r only gives a probability statement about the x,y relationship, while the regression equation describes the relationship quantitatively. As a result, regression analysis is often more useful than correlation analysis.

Linear regression is often used to determine calibration curves. In those applications we need to know the confidence interval for both the slope and the y-intercept. Variation in the data lead to two types of uncertainty in the regression line: transla-

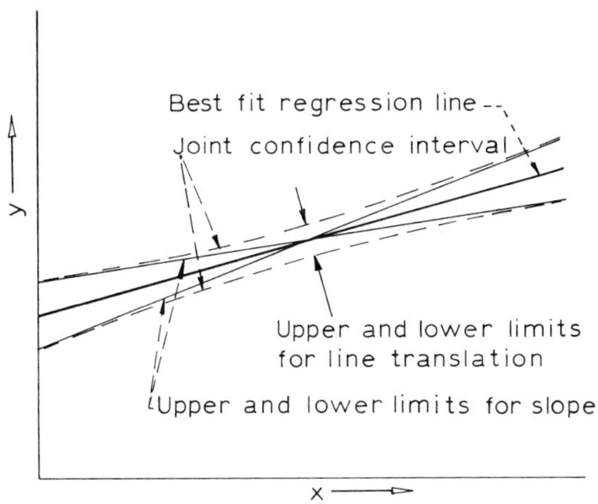

Figure 7-5 Joint confidence interval for a regression line

tion and rotation. This combination results in a band of uncertainty that is narrow in the middle and wide at the ends of the data, as shown in figure 7-5. The width of the band depends on the level of confidence that is used to calculate it.

Another form of error is curvature. When the data is plotted it may be obvious that a curve would fit the line better than a straight line. If multiple values of y are determined for each value of x, i. e. replicate measurements, the variation in y can be used to estimate whether the nonlinearity is significant. This is done by checking whether curvature is significant compared to measurement error. The reduced sum of squared deviations due to curvature is typically calculated as a residual. First, the total sum of squares, SST, is calculated, then subtract the SS due to the regression and the error SS due to the differences between multiple values of y for a given x. The ratio of the residual SS and the error SS is then compared to the F-value from the F-table, see appendix I.

In most cases the presence of nonlinearity becomes quite clear in plotting the data. If the deviation from linearity is significant we have to resort to polynomial regression. Nowadays, many graphics software packages for personal computers offer linear regression capability, in fact, polynomial regression is often available as well. This makes regression analysis very easy to use. The user only has to enter the data, select the appropriate plot, and select the order of the regression. An example is shown in the following table showing data on extruder output versus screw speed for a 1.5″ extruder, see table 7-8.

TABLE 7-8. DATA OF SCREW SPEED VERSUS EXTRUDER OUTPUT

Screw speed [rpm]	Extruder output [gr/min]
20	68.5
40	138.7
60	217.6
80	289.1
100	356.1

The plot of the data is shown in figure 7-6.

It is clear from this figure that there is a very good fit with a straight line. This is confirmed by the value of the coefficient of determination, $r^2 = 0.999$. The equation from linear regression analysis is listed in the top left corner of the plot. Thus, regression analysis becomes very accessible and very easy to use. The user does not have to make any complicated calculations, the computer does all the calculations very quickly and accurately. The plot above with the regression equation was generated within about two minutes using Cricket Graph on a Apple Macintosh Powerbook 170.

It should be noted, however, that the regression equation should always be checked against the actual data. The quickest way to do this is often a visual determination of how far the actual data points are from the regression line.

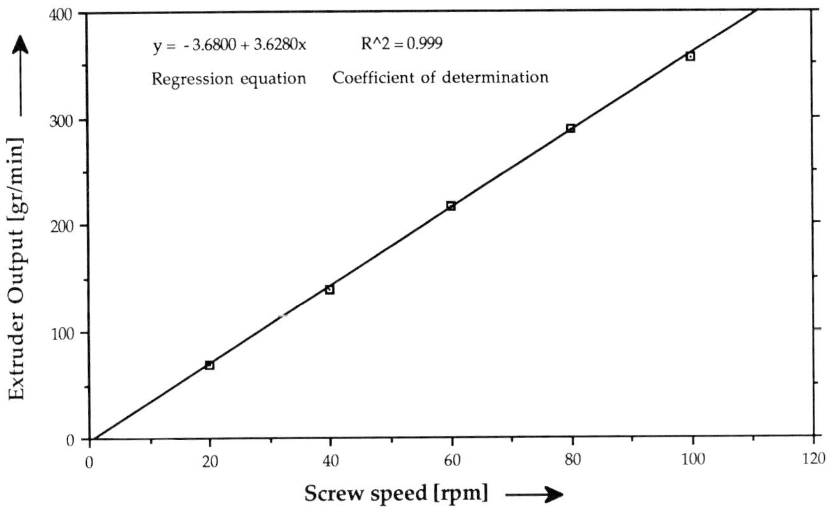

Figure 7-6 Extruder output versus screw speed for 1.5″ extruder

7.3.3 POLYNOMIAL REGRESSION

When data does not fit a straight line, it can often be fitted by a polynomial equation. The general form of the polynomial is.

$$y = b_0 + b_1x + b_2x^2 + b_3x^3 + \ldots\ldots + b_nx^n \tag{7-11}$$

The order of the polynomial is determined by the highest exponent, which is n. A first order polynomial has a highest exponent of only one. A second order or quadratic polynomial has a highest exponent of two, a third order or cubic polynomial a highest exponent of three, etc. As in the linear regression, the regression coefficients, $b_0, b_1, b_2 \ldots b_n$, are determined such that the squared deviations between the points and the line is minimized. This requires solving the following n+1 equations:

$$\begin{bmatrix} N & \Sigma x_i & \Sigma x_i^2 & \cdots & \Sigma x_i^n \\ \Sigma x_i & \Sigma x_i^2 & \Sigma x_i^3 & \cdots & \Sigma x_i^{n+1} \\ \Sigma x_i^2 & \Sigma x_i^3 & \Sigma x_i^4 & \cdots & \Sigma x_i^{n+2} \\ \vdots & & & & \vdots \\ \Sigma x_i^n & \Sigma x_i^{n+1} & \Sigma x_i^{n+2} & \cdots & \Sigma x_i^{2n} \end{bmatrix} \begin{bmatrix} b_0 \\ b_1 \\ b_2 \\ \vdots \\ b_n \end{bmatrix} = \begin{bmatrix} \Sigma y_i \\ \Sigma x_i y_i \\ \Sigma x_i^2 y_i \\ \vdots \\ \Sigma x_i^n y_i \end{bmatrix} \tag{7-12}$$

where N is the number of datapoints.

The order of the polynomial, n, cannot be greater than N-2. If n is greater than N-2, we can calculate an exact solution without using any statistics. Expression 7-12 is written in matrix form, which allows a compact representation of the n+1 equations. Remember that in multiplication of matrices A and B, the element in row i and column j of the product matrix AB is equal to row i of A times column j of B. For instance, if we multiply a nine element (a,b,c, . . . i) matrix with a three element (x,y,z) matrix, we obtain the following result:

$$\begin{bmatrix} a & b & c \\ d & e & f \\ g & h & i \end{bmatrix}\begin{bmatrix} x \\ y \\ z \end{bmatrix} = \begin{bmatrix} ax+by+cz \\ dx+ey+fz \\ gx+hy+iz \end{bmatrix} \tag{7-13}$$

It can easily be verified that for a linear regression with n = 1, the solution of equations 7-12 leads to exactly the same solution as given by equations 7-9 and 7-10. Thus, linear regression is simply a special case of polynomial regression. The advantage being that the equations for linear regression are easy to solve.

If we use a third or higher order polynomial the equations become rather difficult to solve. Particularly in these cases computer programs are quite useful; they can eliminate many tedious calculations. Consider the following data of extruder output versus diehead pressure, see table 7-9.

TABLE 7-9. DATA OF EXTRUDER OUTPUT VERSUS DIEHEAD PRESSURE

Output [gr/min]	Pressure [MPa]
74	9.556
142	11.582
206	13.249
271	14.593
342	15.833

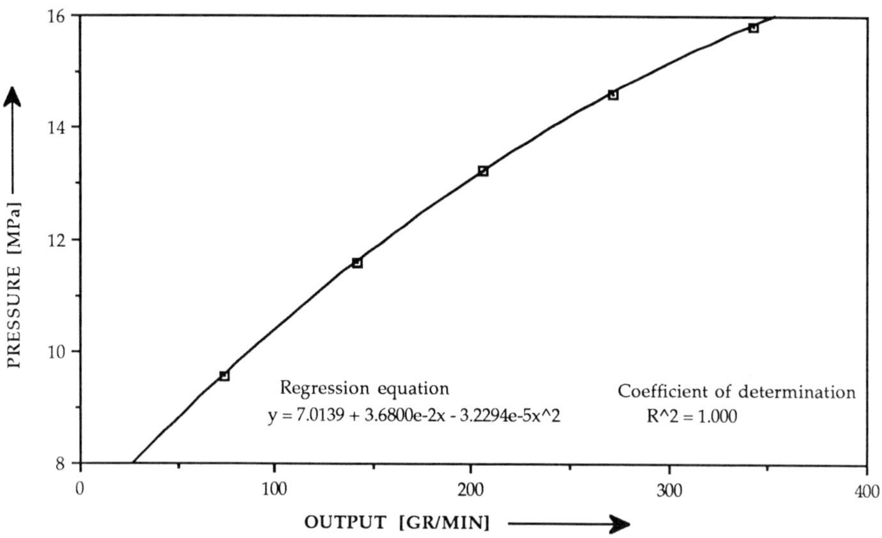

Figure 7-7 Pressure versus output with regression analysis

Using the same graphics program Cricket Graph, we can generate a graph and perform a polynomial regression analysis. The graph is shown in figure 7-7. For this data a second order regression was selected. The curved line drawn through the data points represent the regression equation that is listed in the lower right hand corner of the graph. In this case we obtain a very good fit with a second order polynomial as indicated by the value of the coefficient of determination, $r^2 = 1.000$. As a result of the good correlation obtained, it is not necessary to try a higher order regression. Again, the creation of the graph and the regression analysis took about two minutes on the computer. Doing the same regression analysis by hand would take considerably longer.

The solution of equations 7-12 gives the regression coefficients, however, it does not yield the sums of squared errors needed to evaluate the goodness of the fit. The abbreviated Doolittle (ABDO) algorithm can be used to determine both the regression coefficients and the sums of squared errors. A good discussion on the ABDO method is given by Allen and Cady (22).

Up to this point we have discussed simple situations with only one independent and one dependent variable. In real life problems the situation is often more complicated with two or more independent variables influencing the dependent variable. In such cases we have to use multiple regression.

7.3.4 MULTIPLE REGRESSION

When dependent variable y depends on two or more independent variables the dependence can be represented by a surface. Let us examine the case where y depends on x_1 and x_2 in a linear fashion; the functional relationship can be written as:

$$y = 3x_1 + 2x_2 + 5 \tag{7-14}$$

This relationship can be represented by a three dimensional plot as shown in figure 7-8.

The relationship between y and x_1 and x_2 is represented as a plane in three dimensions. With a linear dependence as shown in equation 7-14 the surface is a flat plane. The more general case with n independent variables will be represented by a surface in n+1 dimensions. Such a surface is called a "response surface." In the simple case shown in figure 7-8 there is no interaction between the independent variable x_1 and x_2. However, interaction effects often occur. In this case the joint effect of the variables is either greater or smaller than the sum of the individual effects. Also, the response surface will no longer be a flat plane but exhibit curvature. When an interaction term is added to equation 7-14 as shown in equation 7-15, the response will show definite curvature.

$$y = 3x_1 + 2x_2 + 0.5x_1x_2 + 5 \tag{7-15}$$

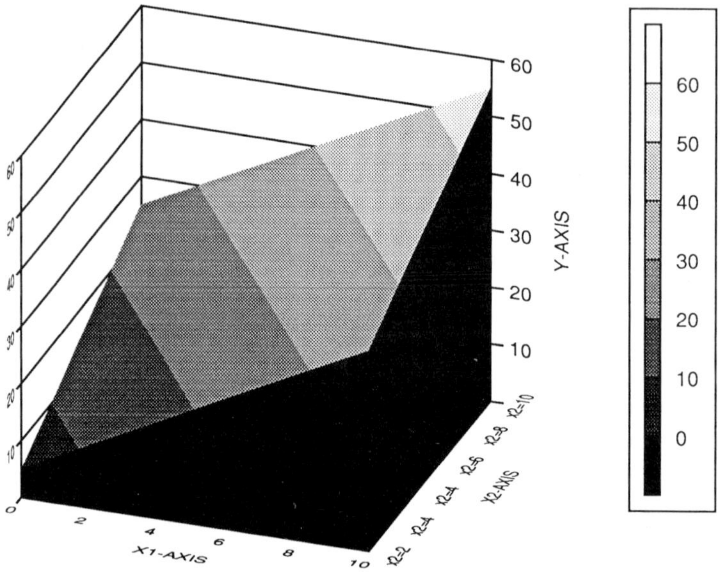

Figure 7-8 Response surface without interactions

Figure 7-9 shows the response surface that goes with equation 7-15. The curvature of the surface is quite apparent.

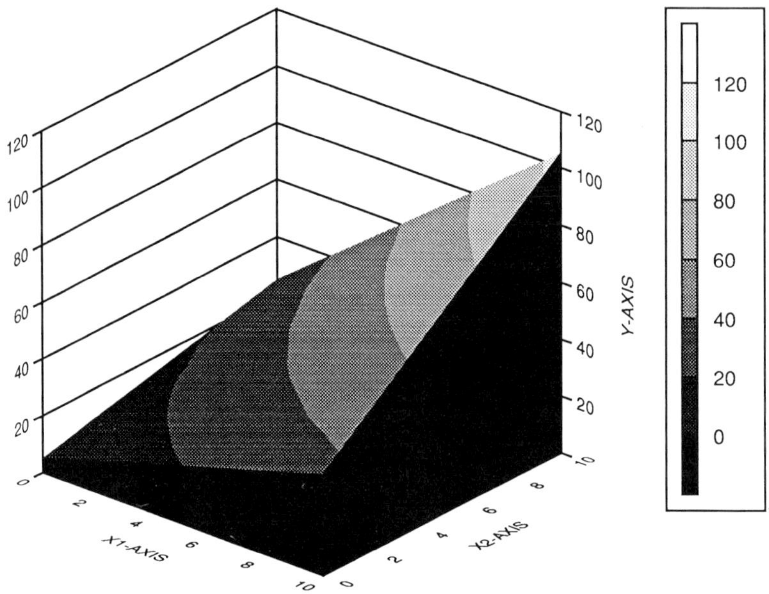

Figure 7-9 Response surface with interaction

We will illustrate multivariable regression by way of an example. It deals with data of extruder output collected at three different screw speeds, at three values of the flight clearance, and at three barrel temperatures. The independent variables are screw speed, x_1, flight clearance, x_2, and barrel temperature, x_3, the dependent variable is extruder output, y. The data is shown in table 7-10; there is a total of 36 data, thus, N = 36.

TABLE 7-10. OUTPUT VERSUS SCREW SPEED, CLEARANCE, AND BARREL TEMPERATURE

x_1 = rpm	x_2 = clearance	$y(x_3 = 175\,°C)$	$y(x_3 = 185\,°C)$	$y(x_3 = 195\,°C)$
20	0.076 mm	74.53 gr/min	72.84	71.08
60	0.076	206.55	205.17	203.93
100	0.076	344.10	342.88	341.39
20	0.15	76.62	73.36	71.09
60	0.15	210.37	207.83	205.18
100	0.15	338.68	336.12	335.02
20	0.25	75.43	72.10	71.08
60	0.25	211.68	208.64	209.16
100	0.25	341.61	337.49	336.41
20	0.51	73.80	69.38	67.32
60	0.51	216.14	212.80	209.57
100	0.51	346.91	341.07	335.80

First, we collect the sums, sums of squares, and sums of products that are required for the regression analysis. This information is listed in table 7-11.

TABLE 7-11. VARIOUS SUMS AND AVERAGES REQUIRED FOR REGRESSION ANALYSIS

$N = 36$	$\bar{x}_2 = 0.2465$	$\Sigma y = 7453.13$
$\Sigma x_1 = 2160$	$\Sigma x_2{}^2 = 3.157884$	$\bar{y} = 207.031389$
$\bar{x}_1 = 60$	$\Sigma x_2 x_3 = 1641.69$	$\Sigma y^2 = 1972494.327$
$\Sigma x_1{}^2 = 168000$	$\Sigma x_2 y = 1840.71112$	$\Sigma x_2{}^2 x_3 = 584.20854$
$\Sigma x_1 x_2 = 532.44$	$\Sigma x_3 = 6660$	$\Sigma x_2 x_3{}^2 = 304304.25$
$\Sigma x_1 x_3 = 399600$	$\bar{x}_3 = 185$	$\Sigma x_2{}^2 x_3{}^2 = 108289.1055$
$\Sigma x_1 y = 541744.19$	$\Sigma x_3{}^2 = 1234500$	$\Sigma x_2 x_3 y = 340349.9233$
$\Sigma x_2 = 8.874$	$\Sigma x_3 y = 1378235.15$	

First, we can check how the variables are correlated. The correlation coefficient for x_1 and x_2 can be determined from the following expression:

$$r_{12} = \frac{N\Sigma x_1 x_2 - (\Sigma x_1)(\Sigma x_2)}{\sqrt{\left[N\Sigma x_1^2 - (\Sigma x_1)^2\right]\left[N\Sigma x_2^2 - (\Sigma x_2)^2\right]}} \qquad (7\text{-}16)$$

If we fill in the numbers from table 7-11 we find that $r_{12} = 0$. This means that x_1 and x_2 are independent or, in other words, orthogonal. In a similar fashion we find that $r_{23} = 0$ and $r_{13} = 0$. Thus, x_1, x_2 and x_3 are mutually independent; this is what we try to achieve when setting up an experiment. In fact, this experiment was designed intentionally to make the x variables orthogonal.

We can now investigate how the x variables correlate with y. First, we will calculate the correlation coefficient for x_1 and y. We can use equation 7-6 or 7-7 for that purpose. Using the numbers in table 7-11 we find for the correlation coefficient for x_1 and y the value $r_1 = 0.73631$. Similarly, we find for x_2 and y that $r_2 = 0.005436$ and for x_3 and y that $r_3 = -0.018499$. Clearly, the screw speed, x_1, has a very strong effect on the output. The effect of flight clearance, x_2, is very small. The effect of temperature, x_3, is also small with a negative correlation, in other words, increasing temperature reduces output.

Since the dominant effect is from the screw speed, we can try a linear regression analysis just with screw speed. The regression coefficients can be calculated using equations 7-9 and 7-10. This leads to $b_1 = 2.4624$ and $b_0 = 59.287$. The simple regression equation thus becomes:

$$y = 59.287 + 2.4624x_1 \tag{7-17}$$

This equation predicts an output at 20 rpm of 108.54 gr/min, at 60 rpm of 207.03 gr/min, and at 100 rpm of 305.53 gr/min. Comparing these predicted values to the actual values shown in table 7-10, it is clear that the predictions at the low and high screw speed are not very good. The prediction at the medium screw speed is reasonable. To determine which effects are significant we have to use multiple analysis of variance.

7.3.5 MULTIPLE ANALYSIS OF VARIANCE

The multiple analysis of variance or MANOVA is used when there are two or more independent variables. The procedure is basically the same as ANOVA, except that there are more factors to take into account. The variances are calculated by deter-

TABLE 7-12. SUMS OF Y VERSUS X_2 AND X_3

	$x_3 = 175$	$x_3 = 185$	$x_3 = 195$	sum
$x_2 = 0.076$	625.18	620.89	616.40	1862.47
$x_2 = 0.15$	625.67	617.31	611.29	1854.27
$x_2 = 0.25$	628.72	618.23	616.65	1863.60
$x_2 = 0.51$	636.85	623.25	612.69	1872.79
sum	2516.42	2479.68	2457.03	7453.13

TABLE 7-13. SUMS OF Y VERSUS X_1 AND X_3

	$x_3 = 175$	$x_3 = 185$	$x_3 = 195$	sum
$x_1 = 20$	300.38	287.68	280.57	868.63
$x_1 = 60$	844.74	834.44	827.84	2507.02
$x_1 = 100$	1371.30	1357.56	1348.62	4077.48
sum	2516.42	2479.68	2457.03	7453.13

mining the means or sums for different groupings of data. The squared deviations of these means or sums from the overall mean ($\bar{y}$), is then used to determine the variances. The various groupings of data with the required sums is shown in table 7-12, 7-13, and 7-14.

TABLE 7-14. SUMS OF Y VERSUS X_1 AND X_2

	$x_2 = 0.076$	$x_2 = 0.15$	$x_2 = 0.25$	$x_2 = 0.51$	sum
$x_1 = 20$	218.45	221.07	218.61	210.50	868.63
$x_1 = 60$	615.65	623.38	629.48	638.51	2507.02
$x_1 = 100$	1028.37	1009.82	1015.51	1023.78	4077.48
sum	1862.47	1854.27	1863.60	1872.79	7453.13

The total sum of the columns and rows should equal the sum of y or Σy. This is a good way to check whether the numbers are correct. The different variances are calculated as shown in the following equations. The total sum of squares is:

$$\Sigma y^2 - (\Sigma y)^2 / N = 1972494.327 - 7453.13^2 / 36 = 429462.4715 \tag{7-18}$$

The sums of squares for x_1 from table 7-13 or 7-14 is:

$$\Sigma(\bar{y}_1 - \bar{y})^2 = (868.63^2 + 2507.02^2 + 4077.48^2) / 12 - 7453.13^2 / 36 = 429094.0206 \tag{7-19}$$

The sums of squares for x_2 from table 7-12 or 7-14 is:

$$\Sigma(\bar{y}_2 - \bar{y})^2 = (1862.47^2 + 1854.27^2 + 1863.60^2 + 1872.79^2) / 9 - 7453.13^2 / 36 = 19.15365 \tag{7-20}$$

The sums of squares for x_3 from table 7-12 or 7-13 is:

$$\Sigma(\bar{y}_3 - \bar{y})^2 = (2516.42^2 + 2479.68^2 + 2457.03^2) / 12 - 7453.13^2 / 36 = 149.72333 \tag{7-21}$$

The sums of squares for x_1x_2 from table 7-14 is:

$$\Sigma(\bar{y}_{12} - \bar{y})^2 = (218.45^2 + 221.07^2 + \ldots 1023.78^2) / 3 - SS_{x_1} - SS_{x_2} - 7453.13^2 / 36 = 164.273 \tag{7-22}$$

where SS_{x_1} and SS_{x_2} are the sums of squares for x_1 and x_2 respectively.
The sums of squares for x_1x_3 from table 7-13 is:

$$\Sigma(\bar{y}_{13} - \bar{y})^2 = (300.38^2 + 287.68^2 + \ldots 1348.62^2) / 4 - SS_{x_1} - SS_{x_3} - 7453.13^2 / 36 = 2.163 \tag{7-23}$$

The sums of squares for x_2x_3 from table 7-12 is:

$$\Sigma(\bar{y}_{23} - \bar{y})^2 = (625.18^2 + 620.89^2 + \ldots 612.69^2) / 3 - SS_{x_2} - SS_{x_3} - 7453.13^2 / 36 = 24.384 \tag{7-24}$$

With all the values of the sums of squares we can construct the MANOVA table:

TABLE 7-15. MANOVA TABLE

Source	deg. freed.	SS	MS	F	p
x_1	2	429094.0206	214547.0103	294101.4535	<0.01
x_2	3	19.15365	6.38455	8.7519534	<0.01
x_3	2	149.72333	74.861665	102.6205	<0.01
x_1x_2	6	164.273	27.378833	37.53096	<0.01
x_1x_3	4	2.163	0.54075	0.74126	
x_2x_3	6	24.384	4.064	5.57094	<0.01
residual	12	8.75392	0.7295		
total	35	429462.4715			

It is abundantly clear in table 7-15 that the main effect on y is from the screw speed, x_1. The second strongest effect is from the barrel temperature, x_3. However, this effect is much smaller than the effect of screw speed. The third largest effect is the interaction between x_1 and x_2. The fourth largest effect is from the flight clearance, x_2. The fifth largest effect is the interaction between x_2 and x_3. The interaction between x_1 and x_3 is not significant.

The next step is to develop the regression equation.

MULTIPLE REGRESSION, EXAMPLE

As a first step we can try a three variable model without interaction terms. The equation can be written as:

$$y = b_0 + b_1x_1 + b_2x_2 + b_3x_3 \tag{7-25}$$

The regression coefficients b can be determined by solving the equations of the following matrix:

$$\begin{bmatrix} N & \Sigma x_1 & \Sigma x_2 & \Sigma x_3 \\ \Sigma x_1 & \Sigma x_1^2 & \Sigma x_1x_2 & \Sigma x_1x_3 \\ \Sigma x_2 & \Sigma x_1x_2 & \Sigma x_2^2 & \Sigma x_2x_3 \\ \Sigma x_3 & \Sigma x_1x_3 & \Sigma x_2x_3 & \Sigma x_3^2 \end{bmatrix} \begin{bmatrix} b_0 \\ b_1 \\ b_2 \\ b_3 \end{bmatrix} = \begin{bmatrix} \Sigma y \\ \Sigma x_1y \\ \Sigma x_2y \\ \Sigma x_3y \end{bmatrix} \tag{7-26}$$

From the MANOVA table we can predict that the b_2 term is going to be insignificant, thus, we can set $b_2 = 0$. For the other coefficients we find $b_0 = 105.06$, $b_1 = 2.4624$, and $b_3 = -0.247458$. At a screw speed of 20 rpm and a temperature of 185 °C, the predicted output becomes y = 108.53 gr/min. At 100 rpm and 185 °C, the predicted output is 305.52 gr/min. Clearly, these predictions are not much better than those obtained with simple linear regression. The advantage of the multiple regression is that we can quantify the influence of temperature, x_3, on the throughput. For example, at a screw speed of 60 rpm the throughput at 175, 185, and 195 °C is 209.50,

207.02, and 204.55 gr/min resp. These predicted values are in reasonable agreement with the actual data.

Working through an example like this with a hand calculator is tedious and time consuming to say the least. Doing it once or twice is good to appreciate the mechanics of the operations involved. However, for repetitive use of regression analysis, particularly multiple regression, the use of computer based software is highly recommended. This will speed up the analysis considerably and will substantially reduce the chance of errors slipping in. Also, it will allow more complex regression with a better chance of obtaining a good fit with the experimental data. In appendix III a partial listing of some commercial software packages for statistical analysis is shown.

7.3.6 STEPWISE REGRESSION

Stepwise regression is a convenient technique in multiple regression. It is best done on a computer and uses the following scheme. First, the factors are ranked by how much they contribute to the variance. Then a regression is done using only the most significant factor and the variances are recalculated. Then the remaining factors are ranked by how much they contribute to the remaining variance, and the most significant factor is then incorporated in the regression equation. Thus, the regression is built up one factor at a time. The process is terminated when incorporation of additional factors does not improve the goodness of the fit. This is determined by the increase in the coefficient of determination, r^2, upon including an additional factor.

7.4 RESPONSE SURFACE METHODOLOGY

Response Surface Methodology, RSM, is an important technique that is used to optimize industrial processes. With RSM, certain important process parameters (independent variables) are varied systematically and the effect on process performance (dependent variable) is measured. An example in extrusion is described by Tadmor and Klein (19). In this example the process parameters varied are screw speed, channel depth of the screw, barrel temperature, flow rate (output), input temperature, and metering section length of the screw, while the effect of these variables on extrudate temperature is measured.

The evaluation of the data involves expressing the performance characteristic, the dependent variable, as a function of the independent variables. The technique that is used for determining the appropriate functional form is regression analysis, see section 7.3. If there are multiple independent variables, the equation describing the independent variable will represent a surface; this is called the response surface. The objective of RSM is to find the point on the surface where the most desirable value of the dependent variable occurs.

The concept of RSM was first developed by Box and Wilson (20). Initially, it was used primarily in the chemical industry, but now it is applied in many different fields. RSM can be considered to consist of two stages.

1. The first order stage with the following steps:

A. A first-order mathematical model is considered,
B. A factorial or other first-order design is performed,
C. The data is fitted,
D. The contours of the response surface are drawn,
E. The direction of the steepest ascent is determined and pursued.

2. The second order stage with the following steps:

A. A second-order mathematical model is considered,
B. A central-composite or other second order design is performed,
C. The data is fitted,
D. The contours are drawn,
E. A canonical analysis is performed and the optimum is located.

An example of a three dimensional plot that is typically developed in RSM is shown in figure 7-1 with a corresponding contour plot shown in figure 7-2. Once such a three dimensional plot is developed, determination of the optimum value is just a question of reading the coordinates of the maximum value of the dependent value.

7.5 EVOLUTIONARY OPERATION, EVOP

In an operating production facility it is often difficult to do experiments by varying several factors over a considerable range, because this can cause loss of control and can be very costly. A useful technique in such cases is evolutionary operation or EVOP. This involves a search for process improvement by making very small changes in the levels of the factors. This technique was developed by Box and Draper (21).

The basic EVOP design is a two-factor, two-level factorial with a center point, as shown in figure 7-10.

The first center point is the normal operating point. Some of the typical steps involved in EVOP are:

- Select two controllable factors,
- Change factors in small steps systematically,

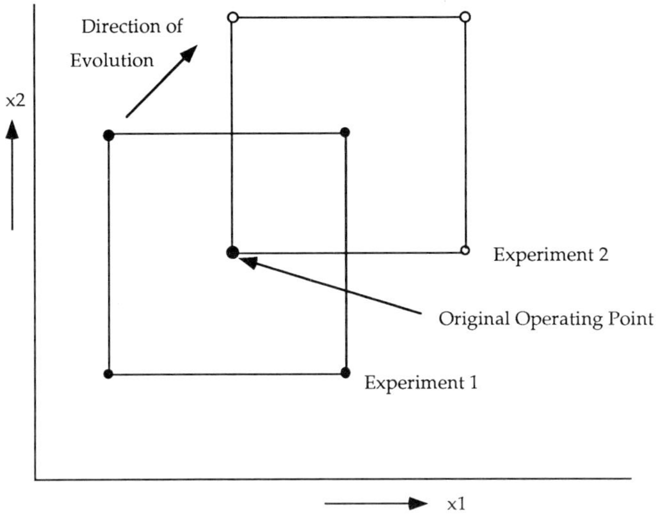

Figure 7-10 Two-variable, two-level factorial with center point

- After a second repetition estimate the effects,
- Move to the indicated better operating point,
- Start new EVOP program.

EVOP is basically a gentle form of response surface methodology. The new operating point is obtained by moving the center point in both directions in proportion to the size of the effects of the two factors. This is the direction of the steepest ascent. The process is continued until no further process improvements can be realised.

7.6 THE TAGUCHI METHOD

The name often associated with design of experiments is that of Dr. Genichi Taguchi. However, Taguchi offers more than techniques of experimental design and analysis. Taguchi has developed a complete and integrated system to develop specifications, engineer the design to these specifications, and manufacture the product to specifications (22).

Quality is defined by Taguchi as the characteristic that avoids loss to society from the time the product is shipped. Quality loss is the financial loss imparted to society after a product is shipped. This loss is measured in monetary units, e.g. dollars, francs, yen, etc., and is related to quantifiable characteristics. This definition goes much beyond the old definition of quality as simply meeting specifications. Two products designed to perform the same function may both meet specifications, but impart different losses to society. Thus, simply meeting specifications is a poor measure of quality.

The quality loss is defined by Taguchi's "loss function." This loss functions relates the financial loss to the functional specification using a quadratic relationship that comes from a Taylor series expansion. The quadratic is of the form of a parabola, see figure 7-11.

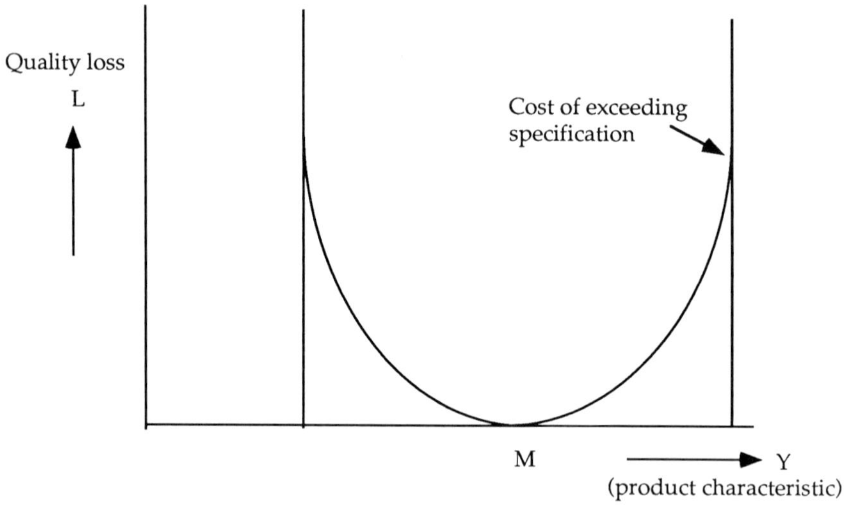

Figure 7-11 The functional form of the Taguchi loss function

The loss function typically is of the following form:

$$L = K(Y-M)^2 \tag{7-27}$$

where L is the loss, K the loss constant, M the mean target value, and Y a particular product characteristic, such as diameter or thickness.

Taguchi has defined several signal to noise ratios as measures of the process variability. In its basic form, the S/N is simply the ratio of the mean to the standard deviation. This is the inverse of the coefficient of variation, which is the standard deviation divided by the mean, as discussed in section 1.3.4. There are three main S/N ratios that can be applied to a wide range of response variables. Type N is for cases where the nominal is best; this typically applies to characteristics such as dimensions, color, resistance, etc. The N type S/N ratio is expressed as:

$$S/N_N = 10\log_{10}\frac{(S_m - V_e)\frac{1}{n}}{V_e} \tag{7-28}$$

where n is the number of observations and y_i an observation.

S_m is the sum of squares due to the mean:

$$S_m = \frac{(\Sigma y_i)^2}{n} \tag{7-29}$$

V_e is the variance:

$$V_e = \frac{\Sigma y_i^2 - \frac{(\Sigma y_i)^2}{n}}{n-1} \tag{7-30}$$

The type S signal to noise ratio is for cases where smaller is better. This applies to characteristics such as contamination, noise, toxicity, etc. It is expressed as:

$$S/N_S = -10\log_{10}\frac{1}{n}(\Sigma y_i^2) \tag{7-31}$$

The type B signal to noise ratio is used in situations where bigger is better. This applies to characteristics such as strength, efficiency, etc. It is expressed as:

$$S/N_B = -10\log_{10}\frac{1}{n}(\Sigma\frac{1}{y_i^2}) \tag{7-32}$$

Taguchi uses two types of experimental design. The first is parameter design and is used to find the levels of the factors that will result in the least variation. The second is tolerance design. This is used only if the variation in product or process is beyond the tolerance limits. Systematic changes are made on a tolerance magnitude to determine which of the factors contribute most to the variation in the end product. For more detailed information the reader is referred to reference 22.

REFERENCES

1. W. Shewhart, "Economic Control of Quality of Product," Van Nostrand Reinhold, New York (1931)
2. K. Ishikawa, "Guide to Quality Control," Nordica International Ltd., Hong Kong (1976)
3. A. C. Duncan, "Quality Control and Industrial Statistics," Richard D. Irwin, Inc., Homewood, IL, Fifth Edition (1986)
4. H. H. Kun et al., "The Measurement Process," National Institute of Standards and Technology Special Publication Number 300, Government Printing Office, Washington (1969)
5. Western Electric, "Statistical Quality Control Handbook," Western Electric Company, Inc., Newark, NJ (1956)
6. "General Motors Statistical Process Control Manual," GM-1693, General Motors Corporation, Detroit, MI (1984)
7. G. Menges and H. Recker, eds., "Automatisierung in der Kunststoffverarbeitung," Hanser Publishers, Munich (1986)
8. A. N. Sahagen, "Silicon-on-Sapphire Pressure Transducer," Measurements & Control, October (1987)
9. C. Rauwendaal, "The Effect of Increased Flight Clearance on Extruder Performance," SPE ANTEC, 108–110 (1989)
10. H. Görman and H. Pütz, "Thermocouples with Increased Accuracy for Plastics Processing Machines," Industrial and Production Engineering, January (1980)
11. H. E. Harris, "A look at the Thinking Behind a New Extruder Temperature Control Approach," Plastics Technology, February, 22–29 (1982)
12. S. S. Wilks, "Mathematical Statistics," John Wiley & Sons, New York, NY (1962)
13. B. Maddock, "A Visual Analysis of Flow and Mixing in Extruder Screws," SPE Journal, 15, 383–389 (1959)
14. C. Goldsberry, "Molders Say Resin Inconsistency Remains a Big Quality Problem," Plastics News, March 4 issue, p. 1 and 16 (1991)
15. S. Basu and C. Rauwendaal, "A Simple test for Sagging Behavior in Blow Molding," 40th SPE ANTEC, San Francisco, CA, 723–726 (1982)
16. C. Rauwendaal, "Polymer Extrusion," Carl Hanser Verlag, Munich, in US: Oxford University Press, 2nd print (1990)
17. ASQC Automotive Division/AIAG (Automotive Industry Action Group), "Measurement Systems Analysis Reference Manual," AIAG (1990)
18. J. M. Juran and F. M. Gryna, "Juran's Quality Control and Handbook, McGraw-Hill, Inc., NY, Fourth Edition (1988)
19. Z. Tadmor and I. Klein, "Engineering Principles of Plasticating Extrusion," Van Nostrand Reinhold, New York (1970)
20. G. E. P. Box and K. B. Wilson, "On the Experimental Attainment of Optimum Conditions," J. Royal Statistical Society, Series B, Vol. 13 (1951)
21. G. E. P. Box and N. R. Draper, "Isn't My Process Too Variable for EVOP?," Technometrics, Volume 10, 439–444 (1969)
22. G. Taguchi, "Introduction to Quality Engineering," Asian Productivity Organization, Tokyo (1986)
23. E. Giese, "Drücke Statisch und Dynamisch Messen bis 600 °C," Kunststoffe, 7, 588–589 (1991)
24. J. A. Clements, "Process Capability Calculations for Non-Normal Distributions," Qualityware (1990)
25. R. L. Plackett and J. P. Burman, "The Design of Optimum Multifactorial Experiments," Biometrika, 33, 305 (1946)

APPENDIX I. F TABLE FOR VALUES SIGNIFICANT AT 1 %

	$n_1 = 1$	2	3	4	5	6	8	10
$n_2 = 2$	98.50	99.00	99.17	99.25	99.30	99.33	99.37	99.40
3	34.12	30.82	29.46	28.71	28.24	27.91	27.35	27.23
4	21.20	18.00	16.69	15.98	15.52	15.21	14.66	14.55
5	16.26	13.27	12.06	11.39	10.97	10.67	10.16	10.05
6	13.75	10.92	9.78	9.15	8.75	8.47	7.98	7.87
7	12.25	9.55	8.45	7.85	7.46	7.19	6.72	6.62
8	11.26	8.65	7.59	7.01	6.63	6.37	5.91	5.81
9	10.56	8.02	6.99	6.42	6.06	5.80	5.35	5.26
10	10.04	7.56	6.55	5.99	5.64	5.39	4.94	4.85
11	9.65	7.21	6.22	5.67	5.32	5.07	4.63	4.54
12	9.33	6.93	5.95	5.41	5.06	4.82	4.39	4.30
13	9.07	6.70	5.74	5.21	4.86	4.62	4.19	4.10
14	8.86	6.51	5.56	5.04	4.69	4.46	4.03	3.94
15	8.68	6.36	5.42	4.89	4.56	4.32	3.89	3.80
16	8.53	6.23	5.29	4.77	4.44	4.20	3.78	3.69
17	8.40	6.11	5.18	4.67	4.34	4.10	3.68	3.59
18	8.29	6.01	5.09	4.58	4.25	4.01	3.60	3.51
19	8.18	5.93	5.01	4.50	4.17	3.94	3.52	3.43
20	8.10	5.85	4.94	4.43	4.10	3.87	3.46	3.37
25	7.77	5.57	4.68	4.18	3.85	3.63	3.32	3.13
30	7.56	5.39	4.51	4.02	3.70	3.46	3.17	2.98
40	7.31	5.18	4.31	3.83	3.51	3.29	2.99	2.80
60	7.08	4.98	4.13	3.65	3.34	3.12	2.82	2.63
120	6.85	4.79	3.95	3.48	3.17	2.96	2.66	2.47
∞	6.63	4.61	3.78	3.32	3.02	2.80	2.51	2.32

The value of n_1 in the table refers to the degrees of freedom in the numerator and n_2 to the degrees of freedom in the denominator.

APPENDIX II. TABLE WITH CRITICAL POINTS FOR THE CORRELATION COEFFICIENT (r)

Degrees of freedom	Probability that $\lvert r \rvert$ greater than r due to chance alone			
	$\alpha = 0.01$	$\alpha = 0.02$	$\alpha = 0.05$	$\alpha = 0.10$
1	0.99999	0.99995	0.9997	0.9877
2	0.9990	0.9800	0.9500	0.9000
3	0.9587	0.9343	0.878	0.805
5	0.875	0.833	0.754	0.669
10	0.708	0.658	0.576	0.497
15	0.606	0.558	0.482	0.412
20	0.537	0.492	0.423	0.360
25	0.487	0.445	0.381	0.323
30	0.449	0.409	0.349	0.296
40	0.393	0.358	0.304	0.257
50	0.354	0.322	0.273	0.231
60	0.325	0.295	0.250	0.211
80	0.283	0.257	0.217	0.183
100	0.254	0.230	0.195	0.164

The degrees of freedom is n-2 for paired samples. The probabilities listed are for two-tailed tests, considering both positive and negative values. For a one-tailed test half the probability value should be used.

APPENDIX III. SEVERAL SOFTWARE PACKAGES FOR STATISTICAL ANALYSIS

Statistica for IBM PC and Mac, does multiple regression, general ANCOVA/MANCOVA, quality control, process analysis, experimental design, and more.
By Statsoft, 2325 E. 13th Street, Tulsa, OK 74104

Statgraphics for IBM, does ANOVA, regression analysis, quality control, experimental design, and more.
By STSC, Inc., 2115 East Jefferson Street, Rockville, MD 20852

Stat Process Control for Mac facilitates the process of collecting, organizing, displaying, analyzing, and charting of time-based data.
By Blackhawk Technology, P. O. Box 2013, Morgan Hill, CA 95038-2013

NWA Quality Analyst does standard statistical control charting with process capability.
By Northwest Analytical, 520 N. W. Davis Street, Portland, OR 97209

S-Plus Version 3.0 for Unix and DOS systems with at least 386 processor and 387 math coprocessor, has over 1200 functions for performing exploratory data and statistical analysis, includes ANOVA, linear and nonlinear models, stepwise regression, etc.
By Statistical Sciences, 1700 Westlake Avenue, Suite 500, Seattle, WA 98109

Systat 5.0 for IBM and Mac has statistics, graphics, and data management capability. Statistics include linear, polynomial, stepwise, weighted regression with extended diagnostics, ANOVA, ANCOVA, MANOVA, etc.
By Systat, Inc., 1800 Sherman Avenue, Evanston, IL 60201-3793

Design-Ease Version 2.0, DOS based, helps experimenters improve their product or process. It includes two-level factorial, fractional factorial, and Plackett-Burman designs. Also offered Design-Expert, which draws response surface maps that lead directly to optimum performance.
By Stat-Ease, Inc., 2021 East Hennepin Avenue, Minneapolis, MN 55413-2723

Minitab Statistical Software for IBM and Mac is used for organizing, analyzing, and reporting statistical data; it does SPC, simple and multiple regression analysis, design of experiment, and more.
By Minitab, Inc., 3081 Enterprise Drive, State College, PA 16801

QC Manager is a PC-based laboratory quality control management data evaluation and presentation software, includes Shewhart control charts capability.
By Telecation, Inc., 19423 North Turkey Creek, Suite C, Morrison, CO 80465

JMP Statistical Discovery Software for Mac does multivariate tests, Shewhart control charts, non-linear fitting, and more. Packages for IBM compatibles and other computer systems also available.
By SAS Institute, JMP Sales Dept., SAS Campus Drive, Cary, NC 27330

BMDP/386 for 386 or 486 PCs does stepwise regression, spectral analysis, Box-Jenkins time series, multivariate ANOVA and ANCOVA, and more.
By BMDP Statistical Software, Inc., 1440 Sepulveda Blvd., Los Angeles, CA 90025

Applied Stats Version 3.0 for the Mac; version for IBM also available. Software is used for data acquisition, both variables and attributes data, SPC functions: variables attributes, and short runs, scatter plots, capability indices, histograms, Pareto charts, gauge capability studies, etc.
By Applied Statistics, Inc., 3080 Centerville Road, St. Paul, MN 55117

Statview for Mac is a comprehensive statistical analysis and presentation graphing program. It performs polynomial and stepwise regression, one to sixteen factor ANOVA, complete non-parametric tests, etc.
By Abacus Concepts, Inc., 1984 Bonita Avenue, Berkeley, CA 94704

PCCTK for IBM and Mac is a graphics and statistical software package that provides a variety of control charts and graphs based on Deming's philosophy of statistical quality control systems. Also available **Lab Partner** a software package that offers a systematic approach to experimental design based on the techniques of Dr. Genichi Taguchi.
By Sof-Ware Tools, P. O. Box 8751, Boise, ID 83707

Genesis for IBM is a data acquisition and control software package. It transforms the IBM PC into a multi-function workstation for creating, simulating, and executing real-time data acquisition and process control strategies. It can implement various system architectures including Monitoring, Supervisory Control, and Direct Digital Control.
By Iconics, Suite 110, 132 Central Street, Foxborough, MA 02035

APPENDIX IV. EXPRESSIONS FOR THE NORMAL DISTRIBUTION CURVE AND Z-TABLE

The equation for the standard normal distribution curve is, see figure AV-1:

$$P(x) = \frac{\exp[-(x-\mu)^2 / (2\sigma)^2]}{\sigma\sqrt{2\pi}} \qquad \text{(AV-1)}$$

where μ is the mean and σ the standard deviation of the normal distribution.
The chance that a measurement is larger than $x=\mu+z\sigma$ is given by:

$$F(x \geq \mu + z\sigma) = \int_{\mu+z\sigma}^{\infty} P(x)dx \qquad \text{(AV-2)}$$

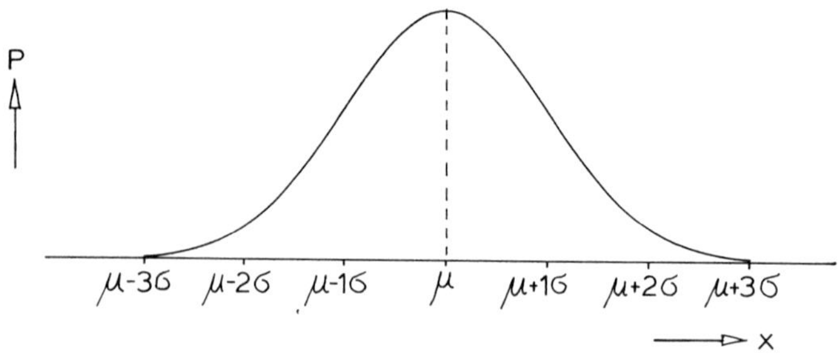

This is the area under the P(x) curve from $x=\mu+z\sigma$ to $x=\infty$ (infinity). Since P(x) is symmetric about $x=\mu$, $F(x\geq\mu+z\sigma)$ is equal to $F(x\leq\mu-z\sigma)$ or:

$$\int_{\mu+z\sigma}^{\infty} P(x)dx = \int_{-\infty}^{\mu-z\sigma} P(x)dx \tag{AV-3}$$

The total area under the curve is 1, thus:

$$\int_{-\infty}^{\infty} P(x)dx = 1 \tag{AV-4}$$

Data on $F(x\geq\mu+z\sigma)$ or $F(x=\mu$ to $x=\mu+z\sigma)$ is often presented in tabular form; an example is the z-table, where $z=(x-\mu)/\sigma$. The table shows the area under the normal curve from z=0 to z or F(z=0 to z), which the same as $F(x=\mu$ to $x=\mu+z\sigma)$, see also figure AV-2.

Z-TABLE. AREA UNDER THE NORMAL DISTRIBUTION CURVE FROM 0 TO Z

Normal Deviate z	0.00	0.01	0.02	0.03	0.04	0.05	0.06	0.07	0.08	0.09
0.0	0.0000	0.0040	0.0080	0.0120	0.0160	0.0199	0.0239	0.0279	0.0319	0.0359
0.1	0.0398	0.0438	0.0478	0.0517	0.0557	0.0596	0.0636	0.0675	0.0714	0.0753
0.2	0.0793	0.0832	0.0871	0.0910	0.0948	0.0987	0.1026	0.1064	0.1103	0.1141
0.3	0.1179	0.1217	0.1255	0.1293	0.1331	0.1368	0.1406	0.1443	0.1480	0.1517
0.4	0.1554	0.1591	0.1628	0.1664	0.1700	0.1736	0.1772	0.1808	0.1844	0.1879
0.5	0.1915	0.1950	0.1985	0.2019	0.2054	0.2088	0.2123	0.2157	0.2190	0.2224
0.6	0.2257	0.2291	0.2324	0.2357	0.2389	0.2422	0.2454	0.2486	0.2518	0.2549
0.7	0.2580	0.2612	0.2642	0.2673	0.2704	0.2734	0.2764	0.2794	0.2823	0.2852
0.8	0.2881	0.2910	0.2939	0.2967	0.2995	0.3023	0.3051	0.3078	0.3106	0.3133
0.9	0.3159	0.3186	0.3212	0.3238	0.3264	0.3289	0.3315	0.3340	0.3365	0.3389
1.0	0.3413	0.3438	0.3461	0.3485	0.3508	0.3531	0.3554	0.3577	0.3599	0.3621
1.1	0.3643	0.3665	0.3686	0.3708	0.3729	0.3749	0.3770	0.3790	0.3810	0.3830
1.2	0.3849	0.3869	0.3888	0.3907	0.3925	0.3944	0.3962	0.3980	0.3997	0.4015
1.3	0.4032	0.4049	0.4066	0.4082	0.4099	0.4115	0.4131	0.4147	0.4162	0.4177
1.4	0.4192	0.4207	0.4222	0.4236	0.4251	0.4265	0.4279	0.4292	0.4306	0.4319
1.5	0.4332	0.4345	0.4357	0.4370	0.4382	0.4394	0.4406	0.4418	0.4429	0.4441

Normal Deviate z	0.00	0.01	0.02	0.03	0.04	0.05	0.06	0.07	0.08	0.09
1.6	0.4452	0.4463	0.4474	0.4484	0.4495	0.4505	0.4515	0.4525	0.4535	0.4545
1.7	0.4554	0.4564	0.4573	0.4582	0.4591	0.4599	0.4608	0.4616	0.4625	0.4633
1.8	0.4641	0.4649	0.4656	0.4664	0.4671	0.4678	0.4686	0.4693	0.4699	0.4706
1.9	0.4713	0.4719	0.4726	0.4732	0.4738	0.4744	0.4750	0.4756	0.4761	0.4767
2.0	0.4772	0.4778	0.4783	0.4788	0.4793	0.4798	0.4803	0.4808	0.4812	0.4817
2.1	0.4821	0.4826	0.4830	0.4834	0.4838	0.4842	0.4846	0.4850	0.4854	0.4857
2.2	0.4861	0.4864	0.4868	.04871	0.4875	0,4878	0.4881	0.4884	0.4887	0.4890
2.3	0.4893	0.4896	0.4898	0.4901	0.4904	0.4906	0.4909	0.4911	0.4913	0.4916
2.4	0.4918	0.4920	0.4922	0.4925	0.4927	0.4929	0.4931	0.4932	0.4934	0.4936
2.5	0.4938	0.4940	0.4941	0.4943	0.4945	0.4946	0.4948	0.4949	0.4951	0.4952
2.6	0.4953	0.4955	0.4956	0.4957	0.4959	0.4960	0.4961	0.4962	0.4963	0.4964
2.7	0.4965	0.4966	0.4967	0.4968	0.4969	0.4970	0.4971	0.4972	0.4973	0.4974
2.8	0.4974	0.4975	0.4976	0.4977	0.4977	0.4978	0.4979	0.4979	0.4980	0.4981
2.9	0.4981	0.4982	0.4982	0.4983	0.4984	0.4984	0.4985	0.4985	0.4986	0.4986
3.0	0.49865	0.4987	0.4987	0.4988	0.4988	0.4989	0.4989	0.4989	0.4990	0.4990
4.0	0.49997									

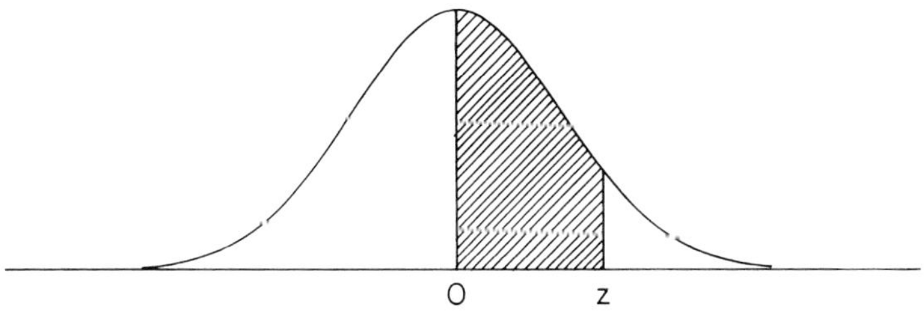

APPENDIX V. GLOSSARY

Accuracy. The closeness of agreement between an observed value and an accepted reference value.

Advanced Statistical Methods. More sophisticated techniques of statistical process analysis and control than included in basic statistical methods; this can include more advanced control chart techniques, regression analysis, design of experiments, advanced problem solving techniques, etc.

Assignable cause. A significant source of variation that is not inherent to the process but has an identifiably reason. Also called special causes, sporadic causes, chaotic causes, and unnatural causes.

Attribute. A product characteristic than can only be described by one of two possible conditions.

Auto-Tuning Controllers. Controllers that can automatically determine and set the appropriate values of the tuning parameters of the controller.

Average. See mean.

Basic Statistical Methods. Applies the theory of variation through use of basic problem-solving techniques and statistical process control; includes control chart construction and interpretation for both variables and attributes data and capability analysis.

Bias. The difference between the true value and the average of repetitive measurements made by a measurement process.

Binomial Distribution. A discrete probability distribution for attributes data that applies to conforming and nonconforming units and underlies the p and np charts.

Capable Process. A process that can deliver in spec. products consistently with a high statistical significance, see also process capability.

Capability Index. The ratio of the tolerance to the process variation; usually the process variation is expressed as 6σ.

Cause and Effect Diagram. A simple tool for individual or group problem-solving that uses graphic description of the various process elements to analyze potential sources of process variation. Also called fishbone diagram (after its shape) or Ishikawa diagram (after its developer).

Cause Listing Diagram. A diagram showing the various possible causes of a problem.

c-Chart. A chart showing the number of defects or nonconformities. It requires constant subgroup size.

Central Limit Theorem. Sample averages (or x-bars) will follow a normal distribution as long as only common cause variations occur.

Central Line. The line on a control chart that represents the average value of the item being plotted.

Central Tendency. The tendency of data to group around a certain value. Measures of central tendency are mean, mode, and median.

Chance Cause. See common cause.

Chaotic Cause. See assignable cause.

Characteristic. A distinguishing feature of a process or its output on which variables or attributes data can be collected.

Chi-square ($\chi^{2)}$ test. Test of normality.

Chronic cause. See common cause.

Cluster. A region in the control chart where points are closely grouped together.

Common Cause. A source of variation that is inherent to the process. Also called random causes, chronic causes, and natural causes.

Consecutive. Units of output produced in succession; a basis for selecting subgroup samples.

Control Chart. A graphical method for determining whether a process is in a state of statistical control.

Control Chart Method. The method of using control charts to determine whether or not processes are in a stable state.

Control Limit. A line on a control chart used as a basis for judging the stability of the process; not to be confused with specification limit.

Correlation Table. A table showing the relationship between two variables.

CP. See process potential index.

CpK. See capability index.

CR. The CR index is the inverse of CP.

CuSum. An advanced statistical method that uses current and recent past process data to detect small to moderate shifts in the process average or variability. CuSum stands for cumulative sum of deviations form the target and puts equal weight on the current and recent past data.

Cycle. A regular, repeating pattern on a control chart characterized by alternating regions of increasing and decreasing values of the data plotted.

Defect. A departure of a quality characteristic from its intended level or state.

Defective Units. A unit of product or service containing at least one defect, or having several imperfections that combined cause the unit not to satisfy intended normal, or reasonably foreseeable usage requirements.

Deviation. The difference between one of a set of numbers and its stated value or intended level.

Devolatilization. The removal of volatile components.

Dispersive mixing. A mixing process in which an intrinsic change takes place in the physical character of one of the components. Agglomerates are reduced in size by fracture due to stresses generated during mixing.

Distribution. A method of describing the variation of a stable system, in which individual values are not predictable but in which the outcome as a group form a pattern that can be described in terms of its location, spread, and shape.

Distributive mixing. Reducing the composition non-uniformity where the ingredients do not exhibit a yield stress.

Event. The occurrence of some attribute.

Fishbone Diagram. See cause and effect diagram.

Freaks. Points located outside the control limits of a control chart.

Gage R&R. The gage repeatability and reproducibility.

Gaussian Distribution. See normal distribution.

GOC. See gravimetric output control.

Gravimetric Output Control. A method of controlling the extruder output by measuring the weight of the material discharged from the feed hopper to the extruder.

Histogram. A bar chart with the height of each bar indicating how many data occur within a certain interval; the width of the bar equals the width of the interval.

Hysteresis. The failure of a property that has been changed to return to its original value when the cause of the change is removed.

Individual. A single unit, or a single measurement of a characteristic, often denoted by the symbol x.

Inspection. The process of measuring, examining, testing, or otherwise comparing the unit with the applicable requirements.

Ishikawa Diagram. See cause and effect diagram.

Item. An object or quantity of material on which a set of observations can be made.

Kurtosis. A characteristic of the shape of a distribution curve where there is a relatively high concentration in the middle and out on the tails with a relatively steep drop in between.

Lower Specification Limit or LSL. The lowest value of a characteristic that the specifications will allow.

LSL. See lower specification limit.

Mean. The sum of the observations divided by the number of observations. Also called average.

Median. It is the middle measurement of an ordered set of numbers with the numbers arranged in order of size. For an even set of numbers, it is the average of the two middle numbers.

Metrology. The science of measurement.

Mixture. A condition of a control chart where there is a relative absence of points near the central line.

Mode. The most frequently occurring value in a set of data.

Moving Range. The difference in value between two or more, most recent, successive samples.

Multimodal Distribution. A distribution with more than one peak.

Natural Cause. See common cause.

Negative Skew. A distribution whose peak is on the upper side of the distribution curve.

Nonconforming Units. Units which do not conform to a specification or other inspection standard; sometimes called discrepant or defective units. The p and np charts are used to analyze systems producing nonconforming units.

Nonconformity. A specific occurrence of a condition which does not conform to a specification or other inspection standard; sometimes called discrepancy or defect.

Non-linearity. The largest deviation of the actual measurements from the straight line characteristic divided by the full scale.

Normal distribution. A bell-shaped distribution described by formula in appendix IV. Also called Gaussian distribution.

np-Chart. A chart for the actual number of defective items in the subgroup. It requires a constant subgroup size.

On/Off Temperature Control. Temperature control where heating or cooling is ap plied either full on or completely off with no intermediate values.

Pareto Chart. A simple tool for problem solving that involves ranking all potential problem areas or sources of variation according to their contribution to cost or total variation.

p-Chart. An attributes control chart for the percentage of defective items in a subgroup when the subgroup size is not constant.

PID Control. See proportional/integral/derivative control.

Plasticating Extruder. An extruder that receives solid material and melts or plasticates the material as it is conveyed to the discharge.

Poisson Distribution. A discrete probability distribution for attributes data that applies to nonconformities and underlies the c and u control charts.

Population. The totality of items or units of material under consideration. Also called universe.

Positive Skew. A distribution whose peak is on the lower side of the distribution curve.

Precision. The extend to which repetitive measurements on a single unit agree.

Prevention. A future-oriented strategy that improves quality and productivity by directing analysis and action toward correcting the process itself.

Probability Plot. A plot of the cumulated distribution on probability paper.

Problem-Solving. The process of determining the causes of problems, as evidenced by certain symptoms, and taking action to eliminate the problem or reduce its severity.

Process. The combination of people, equipment, materials, methods, and environment that produce output - a certain product or service.

Process Analysis Diagram. A diagram showing all the steps in the process.

Process Capability. The ratio of the tolerance (USL-LSL) to the process variation (6σ).

Process Potential Index. Symbolized by CP, is the ratio of the tolerance to 6 times the process standard deviation. It is a measure of spread of the process data.

Proportional/Integral/Derivative Control. Control that uses proportional action combined with integral (reset) and derivative (rate) action.

Quality. The totality of features and characteristics of a product or service that bear on its ability to satisfy a given need.

Random cause. See common cause.

Randomness. A condition in which individual values are not predictable, although they may come from a definable distribution.

Random Sampling. The process of selecting units for a sample of size n in such a manner that all combinations of n units under consideration have an equal chance of being selected as the sample.

Range. The difference between the largest and smallest measurement.

Rational Subgroup. A subgroup gathered in a manner as to give the maximum chance for the measurements in each subgroup to be alike and the maximum chance for the subgroups to differ one from the other.

Repeatability. The variation obtained when one person measures the same quantity several times using the same measuring instrument.

Reproducibility. The variation in measuring averages due to differences among instruments, operators, etc.

Resolution of a Measuring Instrument. The smallest unit of measure which an instrument is capable of indicating.

Run. A consecutive number of points consistently increasing or decreasing or above or below the central line. This can be evidence of the presence of special causes of variation.

Sample. One of several individual pieces or measurements that is collected for analysis. In process control applications, sample is a synonym for subgroup.

Sample Size. The number of units in a sample.

Sampling Interval. In systematic sampling, the fixed interval of time, output, running hours, etc. between samples.

Scatter Diagram. A simple graph showing the cause on the horizontal axis and the effect on the vertical scale, used to assess the scatter in the data.

SEC. See specific energy consumption.

Sensitivity of a Measuring Instrument. The smallest change in the measured quantity which the instrument is capable of detecting.

Shewhart control chart. See $\overline{x}$ and R chart.

Shewhart. Dr. Walter A. Shewhart developed basic methods for statistical monitoring of manufacturing processes.

Sigma (σ). The Greek letter used to designate the standard deviation of a population.

Skew. Asymmetrical.

Skewness. The degree to which a distribution is asymmetrical.

SPC. See Statistical Process Control.

Special cause. See assignable causes.

Specification. The engineering requirement for judging acceptability of a particular characteristic.

Specific Energy Consumption. The amount of energy used per unit volume or unit mass of material.

Sporadic Causes. See assignable causes.

Spread. The span of values from the smallest to the largest in a distribution.

Stability. In process control: the absence of special causes of variation; the state of being in statistical control.

Stability. In measurement analysis: the difference in the average of at least two sets of measurements obtained with the same measuring device on a single unit taken at different times.

Standard Deviation. A measure of variability (dispersion) of observations that is the positive square root of the population variance.

Starve Feeding. A method of feeding an extruder where the feed material is metered in at a rate below the full intake capability of the extruder. In this case, the extruder output is determined by the feeding device, not by the screw speed of the extruder.

Statistic. A quantity calculated from a sample of observations, most often to form an estimate of some population parameter.

Statistical Process Control. The use of statistical methods to monitor and control a process.

Statistics. Deals with the collection, organization, analysis, and interpretation of numerical data.

Stratification. A condition of a control chart where 15 or more points are very close to the central line

Subgroup. One or more events or measurements used to analyze the performance of a process.

Tolerance. The upper limit specification, USL, minus the lower specification limit, LSL.

Tool Control Chart. A control chart that allows current process data to be accurately compared to historical data or to correct data for possible bias.

u-Chart. A chart showing the number of defects or nonconformities. It is used when the size of the subgroup varies.

Unimodal Distribution. A distribution with a single peak.

Universe. See population.

Unnatural Cause. See assignable cause.

Upper Specification Limit. The maximum value of a characteristic that the specifications will allow.

USL. Acronym for upper specification limit.

Variable. A product characteristic that can be one of many possible values.

Variables Data. Quantitative data, where measurements are used for analysis.

Variance. The mean of the squared deviations from the arithmetic mean.

Variation. The inevitable differences among individual outputs of a process; the sources of variation can be grouped into two major classes: common causes and special causes.

Variation Analysis Diagram. A diagram that list the various possible sources of variation.

Weibull distribution. If the life of an item has the density function

$$f(X) = \frac{\beta(X-\gamma)^{\beta-1}\exp[-(X-\gamma)^{\beta/\alpha}]}{\alpha} \quad \text{for} \quad X \geq \gamma$$

f(X) = 0 elsewhere

with $\alpha>0$ and $\beta>0$, then X is said to have a Weibull distribution.

Weld Line. A weak line that can form in a plastic part as a result of insufficient entanglements of the molecules. The can happen when the plastic splits and recombines around a spider support in an extrusion die.

Wilks method. A method of determining process capability for a non-normal distribution.

$\bar{x}$ and R chart. Chart showing the average, $\bar{x}$ and range, R, of groups of data. Also called Shewhart control chart.

x & Rm. Chart showing individual measurements, x, and the moving range, Rm

$\bar{x}$ and s chart. Chart showing the average, $\bar{x}$ and the standard deviation, s.

Zone Analysis. A method of detailed analysis of a Shewhart control chart which divides the x-bar chart between the control limits into three equidistant zones above and below the mean. These zones are sometimes referred to as "sigma" zones.

APPENDIX VI. SOME COMMERCIAL DAS/SPC SYSTEMS FOR EXTRUSION

Although SPC is unquestionably an indispensable tool in extrusion operations, the fact is that there are many extrusion lines without any automatic SPC capability. Even new extrusion systems often do not have adequate SPC capability. The extrusion processor, thus, has to decide whether to upgrade the existing extrusion line or to purchase a new system altogether. As discussed in section 2.7.3, the following capabilities in a DAS are important:

- **Ability to monitor from 40 to 80 variables per machine**
- **Frequency of data collection should be variable**
- **Trending capability on all critical variables**
- **Determination of important statistical parameters in real-time**
- **Alarms for indications of assignable causes of variation**
- **Recipes to store important process variables for various products**
- **Production summaries**

Interestingly, the use of SPC in injection molding seems to have advanced more rapidly than in extrusion. This is somewhat surprising, since the injection molding process is actually more complicated than the extrusion process. One possible reason for this situation may be the fact that the basic operation from one injection molding machine to another is about the same. In extrusion, however, there is a much wider differentiation depending on the product made. For instance, wire coating extrusion is very different from blown film extrusion; both the hardware and the process operations are quite different. The same is true for flat film and sheet extrusion, fiber spinning, profile extrusion, tubing and pipe extrusion, extrusion blow molding, etc.

In retrofitting an existing extrusion line with DAS/SPC capability, there are a number of options. First, if sufficient engineering talent is available in the company, the retrofit can be done in-house. It should be kept in mind, however, that the number of man-hours necessary to do such a project can be quite large.

Second, there are a number of companies that specialize in controller and monitoring systems. Even though these companies usually have more experience with injection molding operations, they can handle extrusion operations as well. A partial listing of these companies is: Allen-Bradley, Barber-Colman, Barco Industries, Camis Systems, Control Process, Data Tech. Research, EPOS, Fisher Controls, Hunkar Laboratories, Mattec, Moog, Plantstar, Production Process, etc. See also Plastics Technology, page 102-111, April, 1990.

The cost of a production monitoring system for a 10-machine plant ranges from about \$30,000 to about \$100,000, depending on the specific features of the system. Some of the important factors to consider when choosing a system are:

- What type of display does the data collection unit have?
- The number of parameters monitored by the data collection unit (DCU).
- The number of machines monitored.
- Third-party interface.
- Is a modem available for customer support?
- Number of computer terminals that have simultaneous access to machine data.
- Can the computer terminal collect data from the machine controllers rather than a DCU?
- Is there a job-scheduling data base?
- Does the same manufacturer make both the system's hardware and software?

A third option is to consider working with a systems house, also called systems integrator. This is basically a company that will analyze your operation and engineer a DAS/SPC system specifically for one or more extrusion lines. Such a company will typically use components from various suppliers and develop the software to make all the components work together as required. There is usually a substantial amount of engineering that has to be done and this bring the cost of this option to a relatively high level. However, for this you will get a custom engineered system.

A fourth option is to use a PC based data acquisition and control system. A example is the Genesis Control Series from Iconics; this system has been used in a number of extrusion operations. The SPC and SQC is an optional addition to Genesis. The SPC on-line features include the ability to view and set parameters, on-line implementation of Shewhart chart trending, a set of off-specification alarms, and real-time statistical data collection to the hard disk for off-line data reduction and analysis. Several other software packages are available, another package that has been used in extrusion is the Camile system from Dow Chemical Company.

A fifth option is to use an integrated, modular control system. An example is the Eurotherm PC3000; this controller incorporates analog and digital control, sequential control, statistical control, and communications. The statistical control features internal calculation of SPC variables, such as standard deviation, mean, upper and lower control limits, etc. and the ability to use this information for real-time SPC/SQC. Various other controllers are available on the market, e.g. the MLS system from Anafaze, Inc., the SIMATIC S5 from Siemens, the MACO 8000 from Barber-Colman, etc.

A sixth option is to work with an extruder manufacturer. Several original equipment manufacturers, OEMs, have developed data acquisition and control systems that have SPC capability. There is a wide range of SPC capabilities in the various systems, however. There are a number of systems that do not offer real-time SPC. This can result in large amount of out of specification product, particularly on large diameter extruders. Some OEM systems will be briefly discussed.

Harrel offers the CC-486 plant-wide control system. This system allows monitoring and control of process parameters on every extrusion line. Multiple terminals can be used. It has multi-tasking capability so that different control and monitoring tasks can be handled simultaneously. The system keeps process information available for analysis for seven days and has full SPC/SQC capabilities. These include x-bar and R charts, histograms, mean, and standard deviation, etc.

Battenfeld Gloucester Engineering manufactures its EXTROL process control system for data collection and control. It is designed to control, monitor, and alarm all important process variables. The data can be graphically trended and each variable is analyzed for its average and standard deviation. Thickness data is also analyzed for process capability. A data link to an IBM compatible PC allows data collected by the EXTROL to be analyzed by any of the software packages available for the PC. A similar system is available from Egan, the Egan system is called CMR. It is a specific extrusion process control system and has little overhead available to perform many statistics. Therefore, the CMR system uses a data link for more detailed statistical analysis.

NRM supplies the Command Sentry 1500 and 2000 Process Control System. The control system monitors and controls all important process variables, provides real-time trending, recipe storage for 120 products, and centralized operation of the entire line. The Sentry Master 3000 Host System is available as an option to allow automated real-time data collection and process and product analysis for one or more Command Sentry 1500 units. Real-time x-bar and R charts are available as an option.

The EPIC system from Davis-Standard/HES is a supervisory process computer. The features of this system include alarm annunciation, recipe storage, production reports, graphic display of historical and real-time trend data, serial communication with auxiliary equipment, and dynamic process graphic color display. The system has some basic SPC features, however, it has no real-time SPC capability.

LIST OF ACRONYMS

ABDO	Abbreviated Doolittle Algorithm
ANOVA	Analysis of variance
CP	Process Potential Index
CpK	Capability Index
CRT	Cathode Ray Tube
CuSum	Cumulative Sum
CV	Coefficient of Variation
DAS	Data Acquisition System
DC	Direct Current
DF	Degrees of Freedom
DOE	Design of Experiments
DSC	Differential Scanning Calorimeter
EVOP	Evolutionary Operation
FS	Full Scale
GOC	Gravimetric Output Control
GR&R	Gage Repeatability and Reproducibility
LCL	Lower Control Limit
L/D	Length to Diameter Ratio
LSL	Lower Specification Limit
MANOVA	Multiple Analysis of Variance
MFI	Melt Flow Index = Melt Index
MI	Melt Index
MPa	Mega Pascal (10^6 Pascal)
MS	Mean Squared Deviation
PDC	Portable Data Collector
PID	Proportional/Integral/Derivative
PSI	Pounds per Square Inch
QC	Quality Control
R&R	Repeatability and Reproducibilty
SEC	Specific Energy Consumption
SPC	Statistical Process Control
S/N	Signal to Noise Ratio
SS	Sums of Squares
SSC	Colums Sums of Squares
SSE	Error Sums of Squares
SSR	Rows Sums of Squares
SST	Total Sums of Squares
TC	Thermocouple
UCL	Upper Control Limit
USL	Upper Specification Limit

NOMENCLATURE

ROMAN CHARACTERS

A_2	constant for determining control limits, see table 3-2
c	number of occurrences of nonconformities in a sample
d_2	constant from table 5-1
D_3	constant for determining control limits, see table 3-2
D_4	constant for determining control limits, see table 3-2
K	number of intervals
L	loss
n	number of production units in the sample
np	number of nonconforming items in a sample of n items
N	number of measurements
M	mean target value
M_i	i-th moment about the mean
p	proportion of nonconforming items in a sample
r	correlation coefficient
r^2	coefficient of determination
R	range
$\bar{R}$	average range
s	sample standard deviation
u	number of occurrences of nonconformities per production unit
v	variance
$\bar{x}$	sample mean
$\tilde{x}$	median
$\hat{x}$	mode
$\bar{\bar{x}}$	grand average

GREEK CHARACTERS

μ	population mean
σ	population standard deviation
χ	see equation 4-5

INDEX